Development of Materials Science in Research and Education (DMSRE30 2020)

IOP Conference Series: Materials Science and Engineering
Volume 1050

Pavlov, Czech Republic
7 - 11 September 2020

Editors:

Zdenek Kozisek
Maria Behulova
Zdenek Potucek

ISBN: 978-1-7138-3336-9
ISSN: 1757-8981

Printed from e-media with permission by:

Curran Associates, Inc.
57 Morehouse Lane
Red Hook, NY 12571

Some format issues inherent in the e-media version may also appear in this print version.

Institute of Physics
Dirac House, Temple Back
Bristol BS1 6BE UK

Phone: 44 1 17 929 7481
Fax: 44 1 17 920 0979

techtracking@iop.org

Additional copies of this publication are available from:

Curran Associates, Inc.
57 Morehouse Lane
Red Hook, NY 12571 USA
Phone: 845-758-0400
Fax: 845-758-2633
Email: curran@proceedings.com
Web: www.proceedings.com

TABLE OF CONTENTS

IOP Conf. Series: Materials Science and Engineering 1050 (2021) 011001 doi:10.1088/1757-899X/1050/1/011001

Proceedings of the 30th DMSRE seminar

Development of Materials Science in Research and Education

Hotel IRIS Pavlov, Czech Republic, 7–11 September 2020

Organized by
Czechoslovak Association for Crystal Growth
Slovak Expert Group of Solid State Chemistry and Physics

Under the auspices of
Institute of Physics of the Czech Academy of Sciences, Prague

Editors
Zdeněk Kožíšek, Mária Behúlová, Zdeněk Potůček

IOP Conf. Series: Materials Science and Engineering 1050 (2021) 011001 doi:10.1088/1757-899X/1050/1/011001

Organizing Committee

Zdenek Kožíšek (conference chair), *Institute of Physics, Czech Academy of Sciences, Prague, Czech Republic*
Aleš Bystřický, *Institute of Physics, Czech Academy of Sciences, Prague, Czech Republic*
Robert Král, *Institute of Physics, Czech Academy of Sciences, Prague, Czech Republic*
Zdenka Poláčková, *Institute of Physics, Czech Academy of Sciences, Prague, Czech Republic*
Petra Zemenová, *Institute of Physics, Czech Academy of Sciences, Prague, Czech Republic*

Scientific Committee

Mária Behúlová, *Faculty of Material Science and Technology SUT, Trnava, Slovak Republic*
Jaromír Drápala, *Technical University of Ostrava, Ostrava, Czech Republic*
Jan Grym, *Institute of Photonics and Electronics, Prague, Czech Republic*
Marian Koman, *Slovak University of Technology in Bratislava, Bratislava, Slovak Republic*
Zdenek Kožíšek, *Institute of Physics, Czech Academy of Sciences, Prague, Czech Republic*
Vladimír Kuchtanin, *Slovak University of Technology in Bratislava, Bratislava, Slovak Republic*
Maroš Martinkovic, *Faculty of Material Science and Technology SUT, Trnava, Slovak Republic*
Petr Mošner, *University of Pardubice, Pardubice, Czech Republic*
Blažena Papánková, *Slovak University of Technology in Bratislava, Bratislava, Slovak Republic*
Jan Polák, *Crytur Turnov, Turnov, Czech Republic*
Zdenek Potůček, *Faculty of Nuclear Sciences and Physical Engineering CTU, Prague, Czech Republic*
Kateřina Rubešová, *University of Chemistry and Technology, Prague, Czech Republic*

IOP Conf. Series: Materials Science and Engineering 1050 (2021) 011001 doi:10.1088/1757-899X/1050/1/011001

Preface

The 30th seminar "Development of Materials Science in Research and Education" (DMSRE30) was held on 7-11 September 2020 in hotel IRIS Pavlov, Czech Republic. The first Joint Seminar in these series was held at Gabčíkovo in 1991 by the initiative of the Czech and Slovak Association for Crystal Growth and the Slovak Expert Group of Solid State Chemistry and Physics. The objective of this meeting offered an opportunity to Czech and Slovak teachers and scientists as well as guests from other countries who are working in the field of Materials Science to present their recent results and experience and to exchange new ideas and information.

The scientific sessions covered the following topics on materials science:

- Trends in the development of materials research
- Education of materials science at the universities
- Information about the research programs of individual institutions
- Information about equipment for preparation and characterization of materials
- Results of materials research

This workshop focused on the creation of a stimulating atmosphere of cooperation and at the support of patient dissemination of scientific ideas and propagation of materials science in education.

In 2020, the coronavirus pandemic influenced the DMSRE30 seminar. Until the last minute, the organizers were not sure whether the DMSRE30 seminar would take place in the form of a personal meeting with on-site presentations or just as a video-conference via the Internet. Finally, we organized both forms of lectures.

At the seminar, 41 participants presented 4 keynote lectures (M. Koman: *Thirty years of seminars "Development of Materials Science in Research and Education"*, M. Buryi: *ZnO nanorods heavily doped with Mo/Er. The effect of post-deposition treatment on defect states and luminescence*, K. Rubešová: *151 years of the periodic table of elements*, L. Koudelka: *Phosphate glasses modified with transition metals*, and Z. Remeš: *The optical emission spectroscopy of the inductively coupled plasma used for modification of chemical, optical and electronic properties of nanostructured ZnO*), 33 on-site talks, and 4 video-conference contributions.

The Proceedings of the DMSRE30 seminar contain 14 contributions.

Z. Kožíšek, M. Behúlová, Z. Potůček (Editors)

IOP Conf. Series: Materials Science and Engineering 1050 (2021) 011002 doi:10.1088/1757-899X/1050/1/011002

Peer review declaration

All papers published in this volume of *IOP Conference Series: Materials Science and Engineering* have been peer reviewed through processes administered by the Editors. Reviews were conducted by expert referees to the professional and scientific standards expected of a proceedings journal published by IOP Publishing.

- **Type of peer review: Single-blind**
- **Conference submission management system: conference Web page https://dms.fzu.cz/30/ and e-mails.**
- **Number of submissions received: 14**
- **Number of submissions sent for review: 14**
- **Number of submissions accepted: 14**
- **Acceptance Rate (Number of Submissions Accepted / Number of Submissions Received X 100): 100**
- **Average number of reviews per paper: 2**
- **Total number of reviewers involved: 24**
- **Any additional info on review process: time delay of review process due to COVID19 pandemic**
- **Contact person for queries: Zdenek Kozisek, E-mail: dms@fzu.cz**

Reference XPS spectra of amino acids

A Artemenko[1], A Shchukarev[2], P Štenclová[1], T Wågberg[3], J Segervald[3], X Jia[3] and A Kromka[1]

[1]Institute of Physics, Academy of Sciences of the Czech Republic, Cukrovarnická 10, Prague 6, Czech Republic
[2]Department of Chemistry, Umeå University, Linnaeus väg 10, Umeå 90736, Sweden
[3]Department of Physics, Umeå University, Linnaeus väg 24, Umeå 90187, Sweden

Corresponding author's e-mail address: artemenko@fzu.cz

Abstract. In this report we present XPS data for five amino acids (AAs) (tryptophan, methionine, glutamine, glutamic acid, and arginine) with different side chain groups measured in solid state (powder form). The theoretically and experimentally obtained chemical structure of AAs are compared. Here, we analyse and discuss C 1 s, N 1 s, O 1s and S 2p core level binding energies, FWHMs, atomic concentrations of the functional groups in AAs. The experimentally obtained and theoretically calculated ratio of atomic concentrations are compared. The zwitterionic nature of methionine and glutamine in solid state was determined from protonated amino groups in N 1s peak and deprotonated carboxylic groups in the C 1s spectrum. The obtained XPS results for AAs well correspond with previously reported data.

1. Introduction

Amino acids (AAs) could be considered as one of the smallest biomolecules which are defined as organic compounds with both an amino group and a carboxyl group in one molecule. The structure of AA includes carboxylic group, an amino group and a side chain. However, the most important is that AAs are the basic building blocks of more complex organic materials such as peptides and proteins which are the objects of numerous investigations in biomaterials science. Thus, it is important to have detailed and reliable experimental data for reference AAs (including the position of core level peaks, characterization of functional groups, atomic concentrations, etc.) for further investigation of more complex organic tissues.

X-ray photoelectron spectroscopy (XPS) is one of the most powerful methods for the characterization of the surface chemistry of a wide variety of materials. XPS is a quantitative non-destructive spectroscopic technique that measures the elemental composition, chemical state and electronic state of the elements that exist within a material. As the depth of analysis in XPS (< 10 nm) is limited by the escape depth of photoelectrons from solid materials it is successfully used as a surface-sensitive method.

In the case of powder samples, like amino acids, care should be taken during the mounting samples for XPS analysis. According to the D. R. Baer [1], the powder samples can be prepared in three ways: (i) by using of the double-sticky tape, (ii) by placing powder in some type of sample holder or containment device, (iii) compressing the powder onto a substrate or sample holder or forming a free-standing pellet. As for AAs, there are a lot of reports dedicated to the investigation of their surface chemical composition by XPS using double-side tape [2], AAs decomposition under irradiation with soft X rays in the gas phase [3] or in a powder form pressed to a uniform layer [4]. Nevertheless, there is a lack of detailed information for high-resolution solid state reference data of AAs in the powder form purchased from various suppliers (including chemical composition and possible contamination).

In this report, we present a detailed XPS analysis of five amino acids (tryptophan, methionine, glutamine, glutamic acid, and arginine) in the solid state which differ in their structure, namely in side chains. Their detailed analysis of chemical atomic composition and high-resolution core levels XPS spectra (C 1s, O 1s, N 1s and S 2p) revealed the differences in theoretical and elemental atomic composition and characteristic functional groups of AAs. The availability of detailed reference data for AAs will be helpful for other researchers in providing of a quality interpretation of obtained XPS spectra.

2. Materials and methods

Different types of AAs in the powder form such as tryptophan (TRP, purity ≥ 98 %), methionine (MET, purity ≥ 98 %), glutamine (GLN, purity ≥ 99 %), glutamic acid (GLU, purity ≥ 99 %), arginine (ARG, purity ≥ 98 %) purchased from Sigma-Aldrich were analyzed by XPS. The chemical structure of AAs is shown in Figure 1. Prior to sample mounting, the sample holder was cleaned in an ultrasonic bath (60 Hz) in ethanol for 15 min and dried in air. The AAs powder was slightly ground immediately before XPS measurements and placed directly on the samples holder by pressing the powder to the surface of the sample holder with a nickel spatula (cleaned with ethanol) by hand. In such a way the pellets of AAs were produced. The sample was pumped in the spectrometer air-lock chamber for 1.5-2 hours before reaching 10^{-7} Torr needed for the transportation of the sample holder into an analytical chamber for XPS measurements.

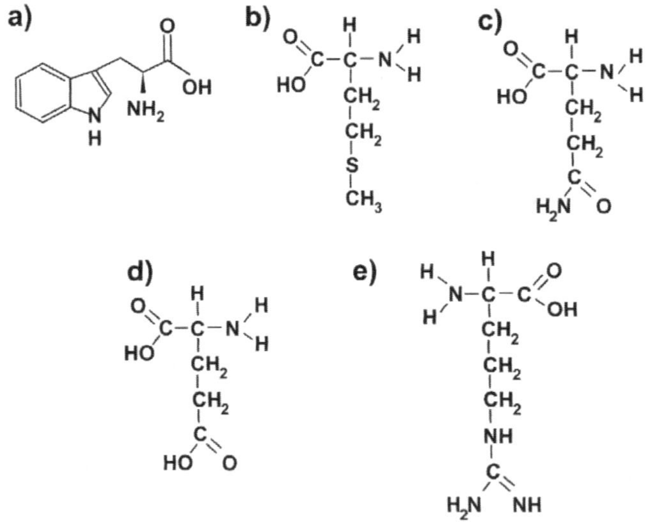

Figure 1. Chemical structure of AAs: a) tryptophan, b) methionine, c) glutamine, d) glutamic acid, e) arginine.

XPS spectra were acquired using Axis Ultra DLD XPS electron spectrometer (Kratos Analytical Ltd., UK) equipped with a monochromated Al Kα X-ray source (1486.6 eV) and hemispherical energy analyzer. The XPS survey spectra were collected at a pass energy of 160 eV at a constant take-off angle of 90°. The high resolution spectra of photoelectron lines were acquired at the pass energy of 20 eV. The charge neutralization system was used during XPS measurements. The Hybrid mode providing the analysis area of 0.3×0.7 mm^2 was applied for collecting XPS spectra. Peak fitting of the measured high resolution spectra was performed by the CasaXPS software using Shirley background and Gaussian/Lorentzian line shapes (GL(30), 70% Gaussian/30% Lorentzian) without fixing peak parameters (binding energies (BE), full width at half maximum (FWHM) and area/intensity). Binding energies (BE) of photoelectron lines were determined with an accuracy of ± 0.1 eV. BE scale was referenced to C 1s component of aliphatic carbon, set at 285.0 eV [2].

3. Results and discussion

3.1. XPS analysis of the total atomic concentration of AAs

The theoretical (calculated from chemical structure shown in Figure 1) and experimental atomic concentrations (AC) for five AAs calculated from core-level XPS peaks are given in Table 1. For measured AAs, except MET, the higher concentration of carbon for experimentally obtained AC can be attributed to adventitious carbon contamination of the samples surface. In the case of MET, the presence of a sulphur signal was detected. The highest experimental AC (~ 30 at. %) of nitrogen was detected on the surface of ARG when other AAs had nitrogen content between 9.8-13.7 at. %. As for oxygen concentration, GLU had 37.6 at. % of oxygen whereas other analyzed AAs had the content of oxygen atoms in the range from 13.4 to 28.1 at. %. XPS analysis of ARG revealed the low chlorine contamination (~ 0.8 at. %) (excluded from experimental AC shown in Table 1). Experimental ACs are in good agreement with theoretical ones within the error of XPS quantification (~ 10% rel.)

Table 1. The theoretical and experimental atomic concentrations of chemical elements calculated from XPS spectra

	Theoretical atomic concentration, at. %				Experimental atomic concentration, at. %			
Amino acid powder	O	C	N	S	O	C	N	S
Tryptophan (TRP)	13.3	73.3	13.3	-	13.4	72.9	13.7	-
Methionine (MET)	22.2	55.6	11.1	11.1	21.3	55.2	10.9	12.6
Glutamine (GLN)	30	50	20	-	28.1	52.9	19	-
Glutamic acid (GLU)	40	50	10	-	37.6	52.6	9.8	-
Arginine (ARG)	16.7	50	33.3	-	17.1	52.5	30.4	-

3.2. XPS data for tryptophan (TRP)

Table 2 and Figure 2 show the C 1s peak of TRP fitted with five components in accordance with its chemical structure (Figure 1). The dominant component at 285.0 eV was attributed to carbon atom bound only to other carbon atoms and hydrogen atoms. The component at 286.2 eV corresponds to sp^2 carbon singly bound to nitrogen, =C-NH. The components at 286.9 eV and 289.1 eV correspond to carbon in H_2N-C-COOH bond and carboxylic group, respectively. The C 1s spectrum of TRP exhibits π-π^* satellite at 291.8 eV. This indicates the presence of aromatic rings, sp^2 state of carbon, in agreement with the chemical structure of TRP (Figure 1). This carbon has BE at 284.7 eV [5]. In spite of the difference between binding energies of C=C and C-(C, H) bonds is 0.3 eV, BE scale was calibrated using C-C/C-H component at 285.0 eV. The obtained atomic concentrations for C-(C, H) – 35.6 at. %, C-N – 30 at. %, H_2N-C-COOH – 6.9 at. %, O=C-OH - 6.1 at. % lead to the calculation of atomic ratio 5.8:4.9:1.1:1 (the resulted atomic ratio is almost the same as the theoretical one).

The O 1s spectrum of TRP was fitted with two components (see Table 2 and Figure 2). The component at 531.9 eV corresponds to doubly bonded oxygen in the carboxylic group (O^{**}=C-OH) when the other component at 532.9 eV was attributed to an oxygen atom in C-OH bond (O=C-O^{**}H) there.

The N 1s spectrum of TRP consists of two components (see Table 2 and Figure 2). The low BE component at 400.5 eV is assigned to nitrogen in C-NH-C group while the second one at 401.9 eV corresponds to the protonated, NH_3^+, amino group. All XPS results are in agreement with data reported in [6, 7].

Table 2. The binding energies (BE), full width at half maximum (FWHM) and atomic concentration (AC) for fitted components of high resolution XPS peaks of TRP

		BE, eV	FWHM, eV	AC, at. %	Chemical bonds
Tryptophan (TRP)	C 1s	285.0	0.85	35.6	**C**-(C,H)
		286.2	0.97	30	=**C**-NH
		286.9	0.97	6.9	H_2N-**C**-COOH
		289.1	0.98	6.1	O=**C**-OH
		291.8	1.60	1.3	π-π^* satellite
	O 1s	531.9	1.15	11.4	O^{**}=C-OH
		532.9	1.52	2.0	O=C-O^{**}H
	N 1s	400.5	0.92	5.9	C-NH-C
		401.9	1.60	7.8	NH_3^+

** chemical state of oxygen.

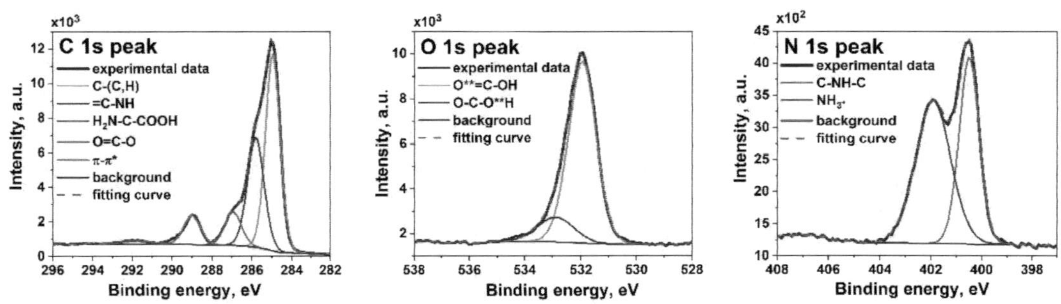

Figure 2. Fitted C 1s, O 1s and N 1s peaks for TRP.

3.3. XPS data for methionine (MET)

The C 1s core level spectrum of MET was fitted with three components corresponding to three chemically different carbon atoms in the molecule (Figure 1). The main component at 285.0 eV corresponds to C-(C, H) bonds; the component at 285.8 eV represents the sum of signals from two carbon atoms bound to the sulphur atom (C-S-C) and amino group (C-NH_2); the component at 288.0 eV arise from carboxylic group (see Table 3 and Figure 3). Table 3 shows the following atomic concentrations for MET: C-(C, H) – 25.7 at. %, C-S-C, C-N – 18.2 at. %, O=C-H – 11.3 at. %. The corresponding atomic ratio for those atomic concentrations is 2.3:1.6:1 which is almost equal to the theoretical value.

The fitting of O 1s core level peak of MET revealed the presence of two contributions. Same as for TRP, oxygen in (O^{**}=C-OH) bond was fitted at 530.9 eV while minor O 1s peak component at 532.0 eV correspond to the O atom in (O=C-O^{**}H) bonds.

The N 1s spectrum of MET demonstrates one component at 401.1 eV which is attributed to protonated NH_3^+ groups. Observation of fully protonated amino group together with dominating O 1s component at 530.9 eV indicates a proton transfer from the carboxylic group of MET to amino group, i.e. zwitterionic form of the molecule in the solid state.

In the case of MET, the S 2p core level spectrum shown in Figure 3 consists of one spin-orbital doublet with BE of S $2p_{3/2}$ components at 163.0 eV corresponding to the sulphur atom and C-S-C bonds. Our XPS data are in agreement with the results presented in [8, 9].

Table 3. The binding energies (BE), full width at half maximum (FWHM) and atomic concentration (AC) for fitted components of high resolution XPS peaks of MET

		BE, eV	FWHM, eV	AC, at. %	Chemical bonds
	C 1s	285.0	0.94	25.7	**C**-(C,H)
		285.8	1.33	18.2	**C**-S-C, **C**-N
Methionine (MET)		288.0	1.04	11.3	O=**C**-OH
	O 1s	530.9	1.19	19.5	O**=C-OH
		532.0	1.21	1.8	O=C-O**H
	N 1s	401.1	1.48	10.9	NH_3^+
	S $2p_{3/2}$	163.0	0.86	12.6	C-S-C

**chemical state of oxygen.

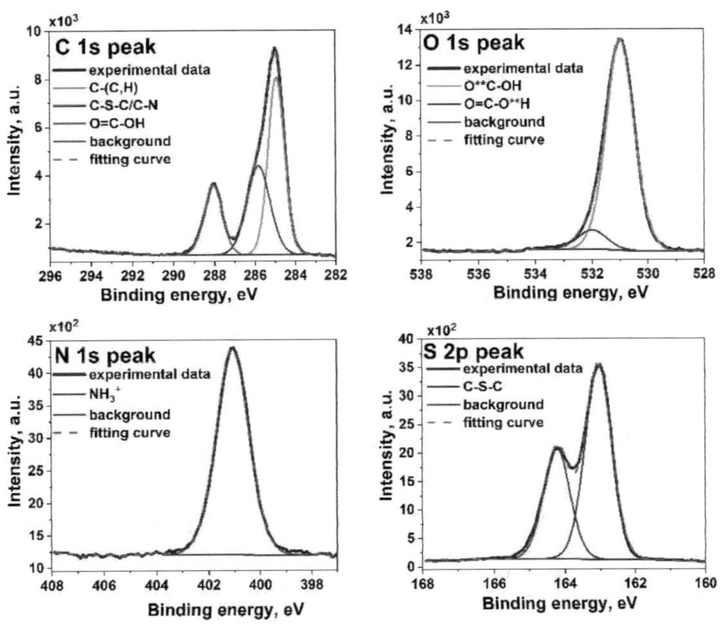

Figure 3. Fitted C 1s, O 1s, N 1s and S 2p peaks for MET.

3.4. XPS data for glutamine (GLN)

According to the chemical structure of GLN (Figure 1, b), there are three chemically different carbon atoms: C-(C, H), $C-NH_2$, and "carboxylic type" carbon - O=C-OH and NH_2-C=O with corresponding atomic ratio 2:1:2. Correspondingly, C 1s core level peak was fitted with three components (see Table 4 and Figure 4). The assignment of chemical bonds is the same as earlier. Observed atomic concentrations for C-(C, H) – 29.1 at. %, C-N – 11.3 at. %, and COOH/NH_2-C=O – 19.7 at. % produce corresponding atomic ratio 2.6:1:1.7 which is close to the theoretical one.

The O 1s spectrum is very similar to ones for TRP and MET, and N 1s spectrum to TRP. Two components in N 1s spectrum indicate that only the "pure" amino group, C-NH_2, is protonated, but not nitrogen in the NH_2-C=O. XPS data for GLN presented above agree with the previous study [2].

Table 4. The binding energies (BE), full width at half maximum (FWHM) and atomic concentration (AC) for fitted components of high resolution XPS peaks of GLN

		BE, eV	FWHM, eV	AC, at. %	Chemical bonds
Glutamine (GLN)	C 1s	285.0	1.10	21.9	**C**-(C,H)
		286.1	1.19	11.3	**C**-N
		288.1	1.14	19.7	O=**C**-O, O=**C**-NH_2
	O 1s	531.2	1.36	24.9	O^{**}=C-OH, O^{**}=C-NH_2
		532.5	1.70	3.2	O=C-O^{**}H
	N 1s	399.6	1.21	8.7	NH_2
		401.2	1.55	10.3	NH_3^+

**chemical state of oxygen.

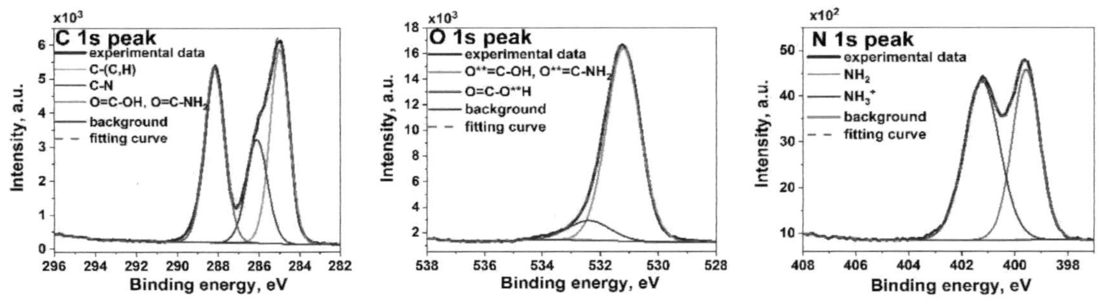

Figure 4. Fitted C 1s, O 1s and N 1s peaks for GLN.

3.5. XPS data for glutamic acid (GLU)

The C 1s core level spectrum of GLU was fitted in accordance with the chemical structure shown in Figure 1 with three components: C-(C, H) bonds at 285.0 eV, C-N bonds at 286.1 eV and O=C-O bonds at 288.6 eV (see Table 5 and Figure 5). The atomic concentrations of chemical bonds in fitted C 1s spectrum of GLU are shown in Table 5. In the case of GLU, the atomic ratio in the C 1s spectrum is 1.6:1:1.6 which is only slightly different from the theoretical ratio 2:1:2.

The O 1s core level spectrum of GLU was fitted with two components same as for TRP: oxygen in (O^{**}=C-OH) bond at 531.6 eV and O atom in (O=C-O^{**}H) bond at 533.0 eV.

In contrast to GLN, the N 1s spectrum of GLU demonstrated only one component at 401.4 eV which was attributed to protonated NH_3^+ groups (see Figure 5). Notice, the presence of a fully protonated amino group together with O 1s component at 531.6 eV revealed a proton transfer from the carboxylic group of GLU to the amino group. Thus, GLU has a zwitterionic form of the molecule in the solid state same as MET. The XPS results obtained for GLU corresponds with previously reported data in [2].

Table 5. The binding energies (BE), full width at half maximum (FWHM) and atomic concentration (AC) for fitted components of high resolution XPS peaks of GLU

		BE, eV	FWHM, eV	AC, at. %	Chemical bonds
Glutamic acid	C 1s	285.0	1.14	20.1	**C**-(C,H)

(GLU)		286.1	1.23	12.4	**C**-N
		288.6	1.54	20.1	O=**C**-OH
	O 1s	531.6	1.59	27.2	O**=C-OH
		533.0	1.70	10.4	O=C-O**H
	N 1s	401.4	1.51	9.8	NH_3^+

**chemical state of oxygen.

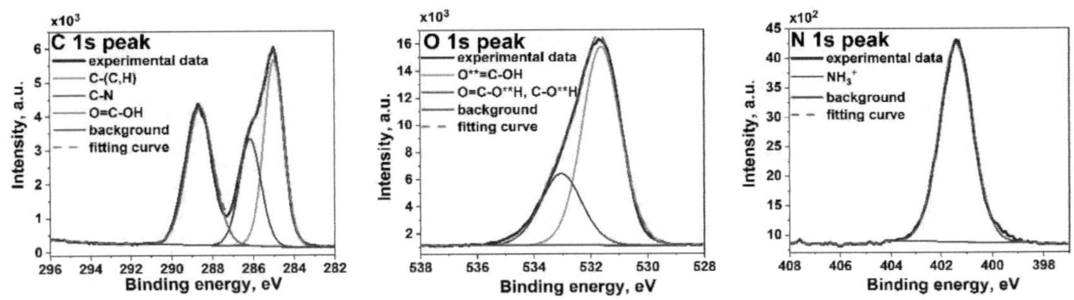

Figure 5. Fitted C 1s, O 1s and N 1s peaks for GLU.

3.6. XPS data for arginine (ARG)

Table 6 and Figure 6 show the C 1s core level spectrum of ARG fitted with four components in accordance with chemical structure (see Figure 1). The bonds at 285.0 and 285.8 eV were attributed to C-(C, H) and C-N bonds, respectively. In contrast to previous AAs, in the C 1s spectrum of ARG the presence of HN=C-(NH, NH_2) bonds results in the spectral component at 287.7. The component at 288.7 eV corresponds to the carboxylic group. Observed atomic concentrations for C-(C, H) – 20.2 at. %, C-N – 15.6 at. %, HN=C-(NH, NH_2) – 7.1 at. % and O=C-OH – 9.6 at. % produce corresponding atomic ratio 2.8:2.2:1:1.3 which is almost the same as theoretical (2:2:1:1).

Same as for MET and GLU, the O 1s core level spectrum of ARG was fitted with two components: doubly bonded oxygen in the carboxylic group (O**=C-OH) at 530.8 eV and the oxygen atom in C-OH bond both in carboxylic groups (O=C-O**H) at 531.9 eV.

The N 1s core level peak of ARG is very similar to ones for MET and GLU. Two components at 399.3 and 400.0 eV were attributed to nitrogen in NH_2 and non-protonated C=NH bonds, and C-NH-C bond together with protonated $C{=}NH_2^+$ group, respectively. The XPS data for ARG are in agreement with results from the previous study [2].

Table 6. The binding energies (BE), full width at half maximum (FWHM) and atomic concentration (AC) for fitted components of high resolution XPS peaks of ARG

		BE, eV	FWHM, eV	AC, at. %	Chemical bonds
Arginine (ARG)	C 1s	285.0	1.16	20.2	**C**-(C,-H)
		285.8	1.24	15.6	**C**-N
		287.7	1.08	7.1	HN=**C**-(NH, NH_2)
		288.7	1.21	9.6	O=**C**-OH
	O 1s	530.8	1.25	15.1	O**=C-OH
		531.9	1.59	2.0	O=C-O**H
	N 1s	399.3	1.33	18.5	NH_2, C=NH

	400.0	1.40	11.9	C-NH-C, $C=NH_2^+$

**chemical state of oxygen.

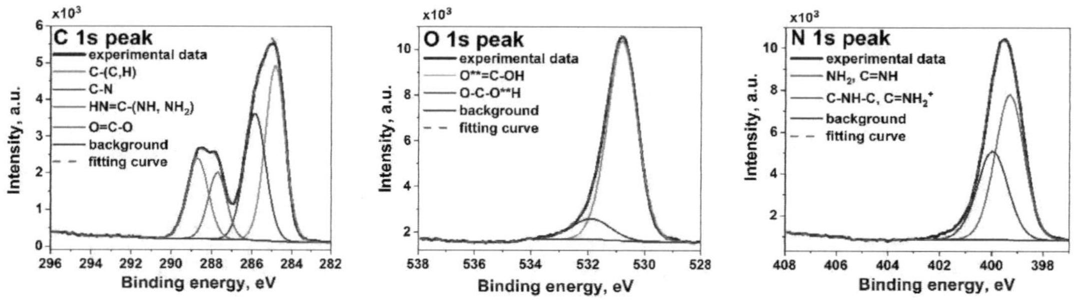

Figure 6. Fitted C 1s, O 1s and N 1s peaks for ARG.

4. Conclusion

We report on a detailed XPS analysis of five AAs in their solid state (powder form) with different side chain groups. Experimentally determined chemical compositions were in good agreement with theoretical values with some possible surface adventitious carbon contaminations. Insignificant chlorine impurity (~ 0.8 at.%) was detected in the case of ARG. High-resolution XPS spectra of C 1s, O 1s, N 1s and S 2p core levels were fitted and analyzed for each of AA in terms of binding energies, FWHMs, atomic concentrations and related to AAs chemical structures. In contrast to other AAs, amino groups of MET and GLU are fully protonated. This makes it possible to conclude that both MET and GLU are present in the zwitterionic form with deprotonated carboxylic groups. We suppose that our XPS data for AAs could be useful for XPS analysis of more complex organic substances (proteins, peptides), biomaterials and tissues. Reference spectra reported in this publication will be deposited in the Open Spectroscopy Database "Spectroscopy Hub" [10].

Acknowledgment

This work was supported by project ESIF and MEYS (Project "FZU- researchers, technical and administrative staff mobility" – CZ.02.2.69/0.0/0.0/18_053/0016627), by the Czech Science Foundation (project No. 20-00925Y) and by CzechNanoLab Research Infrastructure (MEYS CR, no. LM2018110). T. W. acknowledges support from Vetenskapsrådet (2017-04862) Energimyndigheten (45419-1), and Interreg Nord, Region Västerbotten. We also thank the vibrational spectroscopy platform at Umeå University (VISP), the XPS platform and the Nanolab core facility at Umeå University for support.

References

[1] Baer D R 2020 Guide to making XPS measurements on nanoparticles *J. Vac. Sci. Technol. A* **38** 031201
[2] Stevens J S, de Luca A C, Pelendritis M, Terenghi G, Downes S and Schroeder S L M 2013 Quantitative analysis of complex amino acids and RGD peptides by X-ray photoelectron spectroscopy (XPS): Quantitative XPS analysis of amino acids and RGD peptides *Surf. Interface Anal.* **45** 1238–46
[3] Zhang W, Carravetta V, Plekan O, Feyer V, Richter R, Coreno M and Prince K C 2009 Electronic structure of aromatic amino acids studied by soft x-ray spectroscopy *J. Chem. Phys.* **131** 035103

[4] Zubavichus Y, Fuchs O, Weinhardt L, Heske C, Umbach E, Denlinger J D and Grunze M 2004 Soft X-Ray-Induced Decomposition of Amino Acids: An XPS, Mass Spectrometry, and NEXAFS Study *Radiat. Res.* **161** 346–58
[5] Beamson G, Briggs D and SurfaceSpectra Ltd 2000 The XPS of polymers database.
[6] Simões E F C, da Silva J C G E and Leitão J M M 2014 Carbon dots from tryptophan doped glucose for peroxynitrite sensing *Anal. Chim. Acta* **852** 174–80
[7] Barazzouk S and Daneault C 2012 Amino Acid and Peptide Immobilization on Oxidized Nanocellulose: Spectroscopic Characterization *Nanomaterials* **2** 187–205
[8] Thomsen L, Wharmby M T, Riley D P, Held G and Gladys M J 2009 The adsorption and stability of sulfur containing amino acids on Cu{531} *Surf. Sci.* **603** 1253–61
[9] Schiffrin A, Riemann A, Auwarter W, Pennec Y, Weber-Bargioni A, Cvetko D, Cossaro A, Morgante A and Barth J V 2007 Zwitterionic self-assembly of L-methionine nanogratings on the Ag(111) surface *Proc. Natl. Acad. Sci.* **104** 5279–84
[10] Spectroscopy Hub https://spectroscopyhub.com/

IOP Conf. Series: Materials Science and Engineering 1050 (2021) 012002 doi:10.1088/1757-899X/1050/1/012002

ZnO nanorods alloyed with Mo/Er. The effect of post-deposition treatment on defect states and luminescence

M Buryi[1], Z Remeš[1], V Babin[1], V Vaněček[1], K Aubrechtová Dragounová[1], J Mičová[2], L Landová[1] and R Kučerková[1]

[1] Institute of Physics of the Czech Academy of Sciences, Cukrovarnická 10, 162 00 Prague, Czechia
[2] Institute of Chemistry SAS, Dúbravská cesta 9, 845 38 Bratislava, Slovakia

buryi@fzu.cz

Abstract. Free-standing ZnO nanorods alloyed with Er/Mo were synthesized by the hydrothermal growth method. To characterize them, the number of experimental techniques was applied including X-ray diffraction (XRD), scanning emission microscopy (SEM), electron paramagnetic resonance (EPR), photo- and radioluminescence (PL, RL). EPR confirmed the existence of F^+ centres common for ZnO-based structures in the ZnO:Er(30%) nanorods whereas in the ZnO:Mo(30%) this kind of defect was absent. Air annealing at elevated temperatures results in the reduction of F^+ centres in all the materials studied. Moreover, Er^{3+} EPR signal also undergoes changes including broadening in the ZnO:Er. This allowed suggesting oxidation of Er ions on the ZnO nanorods surface. Red luminescence (~680 nm) appears in all studied samples regardless the dopant origin and doping level after the annealing in air. The exciton-related band at 380 nm never observed in the samples before the annealing appears upon the annealing at 350 °C in ZnO:Mo(10%) and ZnO:Er(30%). No such band was observed in the ZnO:Mo(30%) sample under the same conditions. According to SEM there are nanorods no more but microrods upon the content of Mo/Er as compared to the as-grown untreated ZnO as reported in a recent work.

1. Introduction

Scintillators application in computer tomography (CT) and positron emission tomography (PET) is of the utmost importance. This requires the improved coincidence time resolution (CTR) of the scintillating system in the time-of-flight (TOF) PET extension [1]. Significant improvement of the TOF PET is expected at the value around 100 ps [1] what is nowadays the focus of the worldwide research community in a hot field of the ultrafast scintillators engineering. However, the expected limit of the CTR for a scintillating material is 10 ps.

Presently, excellent timing properties have been observed in zinc oxide (ZnO) nanostructures like nanopowders, free-standing (hedgehog) or deposited nanorods. ZnO shows strong room temperature defect related luminescence, high exciton binding energy (60 meV) and most interestingly the ultrafast excitonic luminescence [2]. In most cases it has the hexagonal (wurtzite) structure [3]. Very recently, the preparation of nanocomposite consisting of ultrafast ZnO:Ga scintillating powder incorporated into the scintillating polystyrene matrix was reported [4]. Following the requirements of different applications and band-gap engineering strategy, there were attempts of doping ZnO material (bulk, thin films, quantum dots, and nanocrystalline/nanoparticles) also with other ions originally expected to be paramagnetic (e.g. rare earths or 3d ions like Co^{2+}, Mn^{2+} etc.) [5-8]. Numerous works report data obtained in these systems by EPR and EPR-based techniques (e.g. [9,10]). However, to the best of our knowledge, no works were found reporting studies of the Mo incorporation and its influence on physical and optical properties. Rare earth ions doping of the ZnO matrix is widely known subject, especially from luminescence point of view [11,12]. However, their incorporation inside the host remains rather obscure. Only several works raise this problem by carrying out EPR measurements in ZnO:Er/Gd/Yb [13,14]. Nonetheless, these data are not systematic and can be judged as the preliminary ones gaining very poor insight into the underlying problem. Er^{3+} ions are of great importance because of the luminescence and paramagnetic properties. In most cases EPR spectra measured in as-grown doped or undoped ZnO materials contain only the commonly known resonance line at g factor $g = 1.96$ [15-18] related to an electron captured by an oxygen vacancy, F^+ centre [19,20]. Post-deposition treatment, conditions of growth, additional doping or all this altogether can influence the amount of F^+ centres in the material. Moreover, this also affects the luminescence properties [4,21]. This requires deep understanding of the energy and charge transfer processes including trapping of charge and defects creation. Therefore, this work is aiming at comparison of undoped and doped zinc oxide nanorods alloyed with Er/Mo to find out the peculiarities of the energy and charge transfer and the possibilities of the ZnO-based structures improvement by the embedding of Er or Mo into the structure bulk or their deposition on surface.

2. Experimental conditions

2.1. Samples preparation

All reagents were obtained from commercial suppliers and used without further purification. Zinc nitrate hexahydrate ($Zn(NO_3)_2.6H_2O$) and hexamethylenetetramine (HMTA, $C_6H_{12}N_4$) were purchased from Slavus and erbium nitrate pentahydrate $Er(NO_3)_3.5H_2O$ (Precursor Er) and ammonium heptamolybdate tetrahydrate $(NH_4)_6Mo_7O_{24}.4H_2O$ (Precursor Mo) from Sigma-Aldrich. Deionised water was purified with a So~Safe Water Technologies, having a conductivity 0.20 μS cm^{-1} at 25 °C.

Undoped ZnO nanorods and ZnO nanorods alloyed with Er or Mo with expected nominal compositions $Zn_{1-x}Mo_xO$ ($x = 0, 0.1, 0.3$) and $Zn_{0.7}Er_{0.3}O$ were prepared by the hydrothermal growth method. First, the corresponding stoichiometric amounts of $Zn(NO_3)_2.6H_2O$ and $Er(NO_3)_3.5H_2O$ or $(NH_4)_6Mo_7O_{24}.4H_2O$ were dissolved in 200 ml and 50 ml of deionised water, respectively, with vigorous stirring by a magnetic stirrer. At the same time a 25 mM aqueous solution of HMTA was prepared. All the solutions were filtered through a Whatman 2 filter. Secondly, the solutions of the $Zn(NO_3)_2.6H_2O$ and $Er(NO_3)_3.5H_2O$ or $(NH_4)_6Mo_7O_{24}.4H_2O$ were mixed under vigorous stirring for 15 min. The nominal concentration of each of the $Zn_{1-x}Mo_xO$ ($x = 0, 0.1, 0.3$) and $Zn_{0.7}Er_{0.3}O$ in the final suspensions was 25 mM. The solution of HMTA was added to the mixture. Thirdly, after stirring of the reaction mixture, the hydrothermal growth was performed at 90 °C for 3 hours. Fourthly, the residuals of salts were removed from the grown samples by 5 times washing with deionised water followed by centrifugation at 18 000 rpm (RCF: 23542×g) for 15 minutes. Finally, the samples were dried by lyophilisation.

The MoO_3 nanorods were grown only for comparison with the $Zn_{1-x}Mo_xO$ ($x = 0, 0.1, 0.3$) samples. MoO_3 nanorods were prepared by hydrothermal method as reported in [22]. 1.2 g of $(NH_4)_6Mo_7O_{24}.4H_2O$ was dissolved in 20 ml of deionised water. Subsequently, 65% HNO_3 (80 ml) was added to the solution. Then it was stirred at 85 °C for 30 minutes. After that, the stirring was turned off and the solution held at 85 °C for 1 hour. The synthesized nanorods were isolated from the solution, washed 5 times with deionised water, and dried at room temperature.

2.2. Experimental techniques

XRD was performed using RigakuMiniFlex 600 (Ni-filtered Cu-Kα1,2 radiation) equipped with the NaI:Tl scintillation detector and XRD patterns were compared to the relevant records in the ICDD

PDF-2 database (version 2013). The angular range was 10° – 80° with a step of 0.02° and a scanning speed of 2°/min.

The size and morphology of ZnO nanorods has been checked by scanning electron microscopy (SEM) using MAIA3, TESCAN electron microscope with the in-beam SE detector placed in objective lens and the electron beam energy 5 keV.

The Raman spectra of the undoped ZnO and ZnO:Mo/Er powders were measured at room temperature by a Renishaw In Via Reflex Raman spectrometer using a CCD camera and the 442 nm He-Cd laser excitation. Other parameters were ×100 Olympus objective, 65 μm slits, and a grating with 2400 grooves/mm.

The photoluminescence (PL) measurements were done on powder samples pressed in Suprasil glass tube featuring a low florescence background and converted from wavelength to energy scale taking into account the Jacobian correction [23].

The spectra of steady-state PL excited at the wavelength 340 nm by optically filtered pulsed 1 mW UV LED were recorded in the 355-800 nm spectral range with 2 nm spectral resolution using the spectrally calibrated f/4 double gratings SPEX 1672 monochromator, long-pass filters, a multi-dynode multi-alkali red sensitive cooled photomultiplier, and a lock-in amplifier referenced to the LED frequency 337 Hz [24]. All PL spectra were normalized at the wavelength 355 nm on the same value.

The steady-state PL spectra in the 1200-1650 nm spectral range were recorded with 4 nm spectral resolution under excitation by optically filtered pulsed 150 mW laser diode (LD) at the wavelength 980 nm using the spectrally calibrated Horiba H20IR monochromator, a long-pass filter, an InGaAs photodiode, and a lock-in amplifier referenced to the LD frequency 337 Hz [25].

The optical absorption spectra were measured in the 300-1650 nm spectral range using photothermal deflection spectroscopy (PDS) on pellets with the diameter of 3 mm made from powder pressed onto double sided copper tape glued on copper substrates. The samples were immersed into liquid and the relative temperature of the illuminated sample was measured independently for each wavelength as a deflection of probe laser beam parallel to the sample surface. The spectra were spectrally calibrated by measuring PDS of a black carbon sample and normalized on 1 in the region of full absorption [26].

Ultrafast decays under pulsing X-ray excitation were measured using picosecond (ps) X-ray tube N5084 (Hamamatsu, 40 kV). The X-ray tube is driven by the ps light pulser equipped with a laser diode with the repetition rate up to 1 MHz. The signal was detected by hybrid picosecond photon detector and Fluorohub unit (Horiba Scientific). The setup instrumental response function FWHM is about 76 ps. The spectrally unresolved luminescence decay curves were detected from the surface excited by X-rays.

The radioluminescence (RL) was measured using the Horiba Jobin-Yvon 5000 M spectrometer with Janis liquid nitrogen cryostat and TBX-04 (IBH) photomultiplier operating in the 200–800 nm spectral range. The spectral resolution of the monochromator was 8 nm. All the spectra were corrected for experimental distortions caused by the setup. The powder samples were deposited on a silver paste and pressed on the sample holder. The samples were irradiated by a Seifert X-ray tube operated at 40 kV with a tungsten target. The RL spectra were measured at 300 K and 77 K. PL decay curves were measured using similar setup equipped with ns excitation nanoLED light sources. The deconvolution procedure was applied to the decay curves to determine true decay times (SpectraSolve software package, Ames Photonics).

Electron paramagnetic resonance (EPR) measurements were performed with a commercial Bruker EMXplus spectrometer in the X-band (9.4 GHz) within the 4-296 K temperature range using Oxford Instruments ESR900 cryostat. The spectrometer sensitivity is about 10^{12} spins/mT.

The annealing in air was done in a quartz sample holder within the modular vertical tube furnace Tersid Carbolite with the possibility to elevate temperature up to the 1000 °C.

3. Results and discussion

3.1. Structural and phase purity analysis

In order to check whether the synthesized ZnO:Mo(10%), ZnO:Mo(30%), and ZnO:Er(30%) powders are indeed nanostructured, the samples were examined using SEM. The corresponding images are shown in figure 1. One can immediately see that the undoped ZnO powder (figure 1(a)) represents the commonly known nanorods structure studied in many previous works (see e.g. [17,21,27]) with the circumference of a single nanorod smaller in average than 200 nm having length of the order of 1-2 μm. However, as soon as the doping with 10 at% of Mo was applied, the nanorods size was drastically changed (figure 1(b)). Now it is inappropriate to consider nanorods but rather microrods as the averaged circumference of a single rod has become of the order of 0.5-1 μm with the length over 2 μm. Interestingly, the microrods preserve wurtzite-like [3] hexagonal shape being more massive at the base and getting narrower at the top (see figure 1(b)).

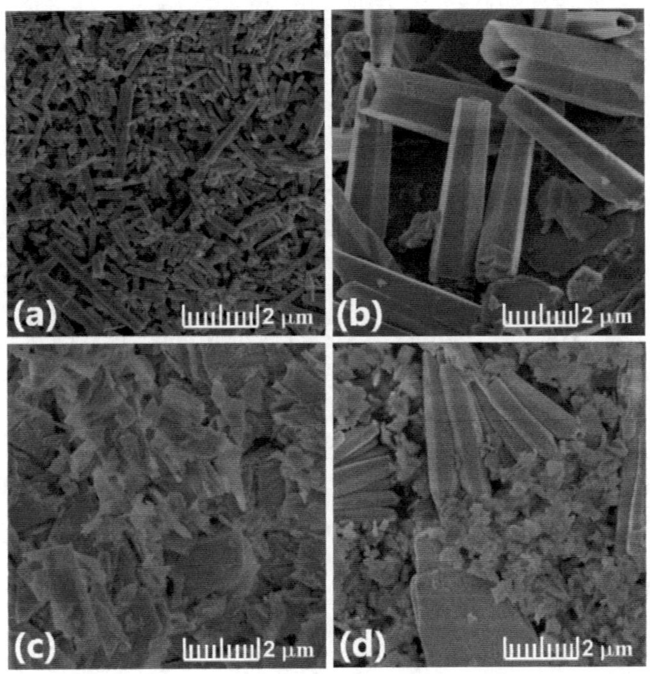

Figure 1. SEM images of the samples: (a) undoped ZnO, (b) ZnO:Mo(10%), (c) ZnO:Mo(30%), (d) ZnO:Er(30%).

The increased level of doping from 10% to 30% results in the degradation of the structures shown in figure 1(b) and so only the shape of flakes can be observed (figure 1(c)). Their shape is no more hexagonal but rather irregular. These flakes are smaller and larger with the size varying from approximately 100×200 nm to 1×2 μm. Doping of the hydrothermally grown ZnO powder with 30% Er results in the appearance of very small flakes approximately 100×200 nm and large hexagonal rods (figure 1(d)) approximately of the size of the microrods observed in the ZnO:Mo(10%) sample (figure 1(b)). Their circumference reaches 0.5-1 μm in average and the length surpasses 2 μm. These rods are also thicker at the base and narrower at the top as well as the ones in the ZnO:Mo(10%).

The XRD patterns of the undoped ZnO and ZnO:Mo/Er samples are shown in figure 2(a). From this follows that the undoped ZnO nanostructures are single phase, composed only of the wurtzite ZnO. As for the ZnO:Mo/Er samples, they exhibit the presence of both the wurtzite ZnO and some other phase(s)/structure(s) further, in general, referred as the not-ZnO_{Mo} and not-ZnO_{Er} in the ZnO:Mo and ZnO:Er, respectively. According to figure 2(a) the wurtzite ZnO phase in the ZnO:Mo(10%) and

ZnO:Er(30%) is dominating. Oppositely, the ZnO:Mo(30%) exhibits much weaker peaks originating from the wurtzite ZnO phase as compared to the not-ZnO_{Mo} phase. In case of the ZnO:Er(30%), the not-ZnO_{Er} phase is unlikely the Precursor Er in its original form as it is too hygroscopic [28]. Probably, the not-ZnO_{Er} phase in the ZnO:Er(30%) sample is the combination of Er-based oxides or hydroxides created out of the part of the Precursor Er decomposition during the hydrothermal reaction. Note, that the not-$ZnO_{Mo/Er}$ reflections emerged in ZnO:Mo/Er samples could not be matched with any relevant record from the ICDD PDF-2 database (version 2013). The not-ZnO_{Mo} reflections are getting stronger in the ZnO:Mo(30%) sample as compared to the ZnO:Mo(10%). Therefore, the not-ZnO_{Mo} phase, most likely, originates from the mix of the residual of Precursor Mo and the products of the partial Precursor Mo decomposition in the ZnO:Mo(10%)/(30%). Note that the Precursor Mo is decomposing upon thermal treatment into MoO_{3-x} oxides or hydroxides as reported in [29]. Therefore, it was impossible to find out precisely the exact origin of the not-$ZnO_{Mo/Er}$ in the ZnO:Mo/Er samples.

To have better insight into the structural and phase properties of the samples, they were examined by Raman spectroscopy. The spectra are shown in figure 2(b). The identification of the peaks originating from the hexagonal wurtzite ZnO phase was done basing on a recent work dedicated to ZnO nanorods (for details see [30]) whereas the groups of peaks belonging to the Precursor Mo were identified by comparing the obtained results (figure 2(b)) with the data reported in [31]. Remarkably, the obtained Raman data (figure 2(b)) are complementary to the XRD data (figure 2(a)) confirming the presence of the Precursor Mo in the ZnO:Mo samples. The undoped ZnO exhibits typical single phase of hexagonal wurtzite structure whereas the Mo doping results in the presence of Precursor Mo phase getting stronger upon the doping level (A and B group of peaks in figure 2(b)). Interestingly, the peaks originating from ZnO are lost in noise in case of ZnO:Mo(30%). On the contrary, it seems that the ZnO:Er(30%) with the same 30% doping level exhibits one relatively weak peak (considering large doping level) which might originate from Precursor Er or rather some hydrate-like modification of it according to [32] besides those originating from ZnO. This allows to conclude along with SEM and XRD data that doping with Er is less harmful to the rods shape and the material phase purity than the doping with Mo.

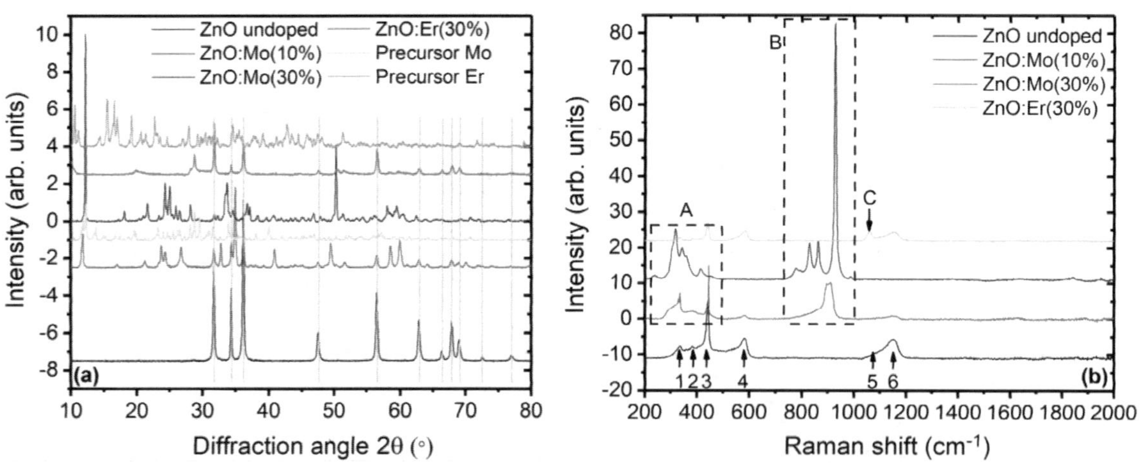

Figure 2. (a) XRD patterns of ZnO:Mo/Er structures prepared by hydrothermal growth along with the precursors used as the source of the specific dopant. Vertical lines indicate the wurtzite ZnO phase according to PDF database record #01-080-0075. (b) Raman spectra of the as-prepared ZnO materials (doping levels are given in a legend). The 1-5 numbers correspond to the specific vibrational modes in the way [30]: 1 - (E_2^{high} - E_2^{low}); 2 - A_1(TO); 3 - E_2^{high}; 4 - A_1(LO); 5, 6 - 2A_1(LO), 2E_1(LO). The group of peaks A and B originate from Precursor Mo [31]. Peak C might originate from Precursor Er or rather a modified Precursor Er-based hydrate.

3.2. Photothermal deflection spectroscopy measurements

By measuring PDS it was possible to determine the optical absorption properties of the ZnO:Mo(10%)/(30%) and ZnO:Er(30%) samples. The corresponding spectra are shown in figure 3. Their envelope function was typical for ZnO [17,33] in undoped ZnO, ZnO:Mo(10%), and ZnO:Er(30%). In ZnO:Mo(30%), the absorption edge was distorted as compared to the rest of samples. Obviously, small content of the ZnO phase confirmed above by XRD and Raman spectroscopy and the presence of parasitic phases are responsible for this. The absorption is peculiar in ZnO:Er(30%). Except the typical for undoped ZnO envelope it exhibits multi-peak structure. These peaks were very narrow. The presence of Er as a dopant allowed to conclude that they can originate from it. Er^{3+} is a rare earth ion with the 4f-4f transitions (shielded from the local crystal (ligand) field). Therefore, comparing with the [34] where the Er^{3+} absorption was measured in germanate glass the typical Er^{3+} absorption transitions were indicated in figure 3.

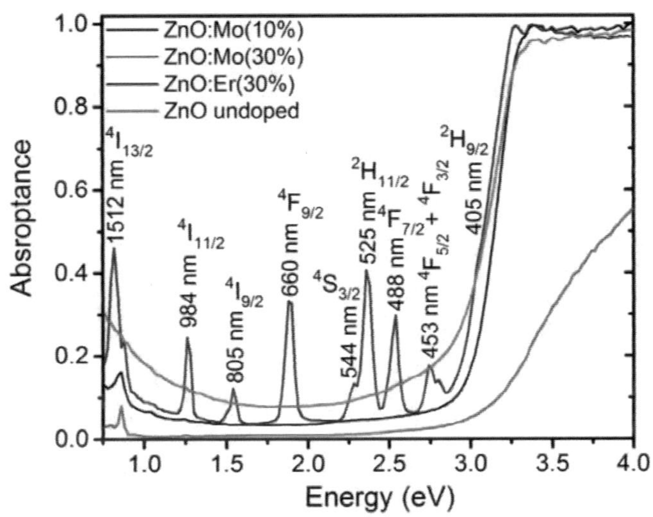

Figure 3. Absorption spectra in undoped ZnO, ZnO:Mo(10%)/(30%), and ZnO:Er(30%). The $^4I_{15/2} \rightarrow {}^{2S+1}X_j$ transitions of Er^{3+} and corresponding wavelengths are indicated.

This is also the reason why it was hard to determine whether the Er^{3+} ions were incorporated into the ZnO microrods, deposited onto the microrods surface or stay as the part of some parasitic phase discussed above in the sample. It is known that the absorption edge in ZnO is shifting to higher energy values upon the increased charge carrier concentration [33,35]. This is indeed observed in figure 3, i.e. the absorption edge in undoped ZnO is shifted to the higher energies as compared to the ZnO:Mo/Er samples. ZnO nanorods have large area and thus, even deposition of the Mo or Er on the rods surface can have the effect of the suppression of the charge carriers concentration. Therefore, one may expect at least deposition of Mo or Er on the wurtzite phase ZnO rods surface. Study of luminescence will complete the obtained results.

3.3. Peculiarities of photoluminescence

PL spectra have been measured in the as-grown ZnO:Mo(10%)/(30%) and ZnO:Er(30%) within the 290 – 4 K temperature range. They are demonstrated in figures 4(a)-(c). At least two bands E1 and E3 peaking at about 1.91 and 3.45 eV could be distinguished in ZnO:Mo(10%). Moreover, the E3 band is further triply split upon decreasing temperature from 290 to 4 K into E3' (3.39 eV), E3'' (3.35 eV), and E3''' (3.18 eV). The PL intensity of the E1 and E3 emission bands is increasing upon the decreasing temperature. The E3', E3'', and E3''' peaks should originate from the bound excitons similar to the ones discussed in [36] for undoped ZnO. Remarkably, the E3' (3.39 eV) and E3'''

(3.18 eV) peaks were also detected in the ZnO:Mo(30%) sample at temperatures below 150 K. The behaviour of the PL intensity seems to be similar to the one observed in ZnO:Mo(10%). Therefore, one may expect influence of Mo ions on ZnO in both cases. However, the E1 band is totally absent at all temperatures in the ZnO:Mo(30%) as can be seen in figure 4(b). The E1 band is present in the PL spectra measured in ZnO:Er(30%), however, shifted to higher energies (1.96 eV). It demonstrates steeper dependence of the PL intensity upon the decreased temperature as compared to the ZnO:Mo(30%) (see figures 4(a),(c)). The E3 band is split triply as well at very low temperatures. The three new peaks 1, 2, and 3 can be observed having maxima at 3.39 eV, 3.33 eV, and 3.25 eV. Remarkably, the maxima of the peaks 1 and E3' and 2 and E3'' are (almost) the same pointing out the possibility that their origin is the same in all three samples with no effect of dopants. However, the PL intensity distribution over the peaks 1 and 2 in the ZnO:Er(30%) and E3' (3.39 eV) and E3'' (3.35 eV) in ZnO:Mo(10%) is different. The peak 2 seems to be unique and thus it was referred to the exciton bound at Er^{3+}. All this may allow to at least expect Mo or Er deposition onto the surface of ZnO microrods, if not the incorporation in the microrods bulk, to affect the energy transfer to excitons.

It is known that annealing in air caused drastic changes to luminescence properties of undoped ZnO nanorods [21]. Therefore, the samples in the present case were also thermally treated in air at elevated annealing temperatures. The corresponding PL spectra were measured at room temperature. They are

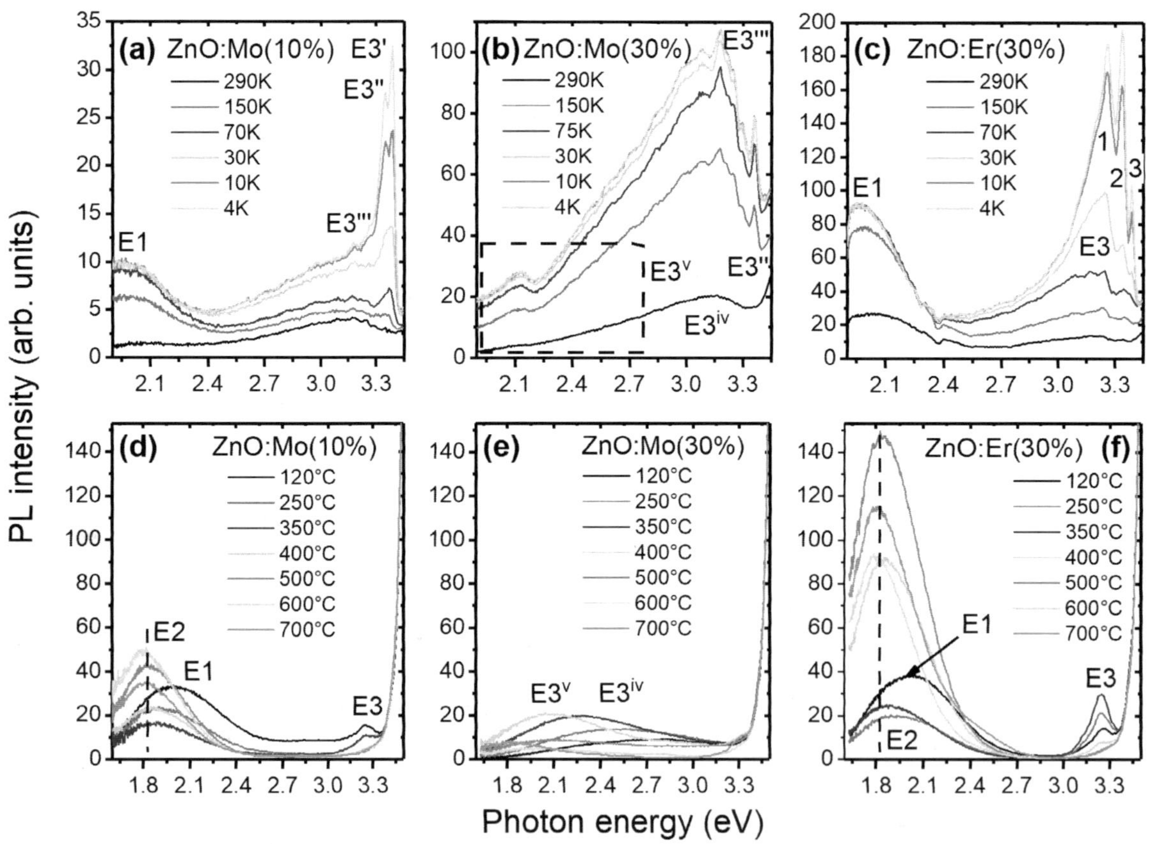

Figure 4. Temperature dependence of PL spectra measured in (a) ZnO:Mo(10%), (b) ZnO:Mo(30%), and (c) ZnO:Er(30%). PL spectra measured in (d) ZnO:Mo(10%), (e) ZnO:Mo(30%), and (f) ZnO:Er(30%) at room temperature upon annealing in air for 15 minutes at temperatures given in legends. E1-E3, E3', E3'', E3''', $E3^{iv}$, and $E3^{v}$ indicate specific emission bands and peaks whereas numbers 1-3 indicate emission of bound excitons in E3 band in the panel (c). The ordinate scale was equalized for the (d)-(f) panels for comparison.

shown in figures 4(d)-(e). The PL spectra of ZnO:Mo(10%) (figures 4(a),(d)) demonstrate E1 and E3 emission after drying at 120 °C. However, as soon as the temperatures is elevated up to 250 °C and higher, the new emission E2 with the maximum at about 1.82 eV (680 nm) appears (to the best of our knowledge it has never been observed before in the ZnO-based structures doped with Mo or Er) while the E1 drops down. Moreover, the E2 is getting stronger upon the annealing up to 600 °C with the small decrease at 350 °C and it goes down again after the annealing at 700 °C.

This trend can be explained by the creation of the new defect state within the ZnO band-gap either on the rods surface or in bulk due to interaction with air. The E3 emission is getting a bit stronger upon the annealing especially at 250-350 °C and disappears completely after 400 °C. Similar effect was observed and discussed in detail in undoped ZnO recently [21].

The situation is even more drastic in ZnO:Mo(30%) sample upon the annealing procedure (figure 4(e)) where the evolution of emission can clearly be observed. After the drying at 120 °C, the $E3^{iv}$ band peaking at ~2.82 eV appears. Its low-energy tail is obviously composed of some other contribution as the annealing at elevated temperatures seemingly causes the band increase and red shift of the maximum to ~2.52 eV for 250 °C, ~2.28 eV for 350 °C, and finally ~2.07 eV for 400 °C (the $E3^{v}$ band in figure 4(e)). After the 400 °C (i.e. annealing at 500-700 °C) the $E3^{v}$ band intensity goes down. This image is very different from what was observed in the ZnO:Mo(10%). Obviously the annealing activates/creates new emission centres, however, most probably, they have nothing in common with that very little of the pure ZnO phase presented in the ZnO:Mo(30%) sample. The E3 band was never observed there.

The effects of much stronger amplitude were observed in the ZnO:Er(30%) sample (figure 4(f)) as compared to the ZnO:Mo(10%) one. The E2 emission band reached maximum at 500 °C and it was about three times of the E2 maximum intensity measured in the ZnO:Mo(10%) after annealing at 400 °C (see figures 4(d),(f)). The E2 exhibited increase all the time within the 250-500 °C with the subsequent drop at 600 °C and increase again at 700 °C. This trend is a bit different than observed in the ZnO:Mo(10%) (figure 4(d)). This may suggest the influence of the dopant on the created defect states. Moreover, since the E2 band was observed in the ZnO:Mo/Er samples, having in common only the ZnO host, one may expect the defect states responsible for the E2 band to be created within the ZnO band-gap, at least at the surface of the microrods. The exciton-related emission in the vicinity of 3.26 eV, the E3 band, is also much stronger in the ZnO:Er(30%) sample. Its intensity reaches maximum after the annealing at 350 °C and then drops down until disappeared completely at 500 °C. It has never appeared again afterwards at higher annealing temperatures. This allows to conclude that the presence of large number of Mo ions results in the exciton emission quenching. Again, the Er doping seems more suitable for the scintillator use purpose. In order to find out the effect of the air annealing on the Er^{3+} emission, ZnO:Er(30%) PL was measured in the infrared range. The corresponding spectra are demonstrated in figure 5.

The observed ${}^4I_{13/2} \rightarrow {}^4I_{15/2}$ optical transition of the Er^{3+} ion is split into several peaks. This is because the ${}^4I_{13/2}$ and ${}^4I_{15/2}$ levels are degenerate, 7 and 8 times, respectively, in a free ion and become split by the local crystal (ligand) field in a host. Therefore, the peaks in figure 5 correspond to the transitions which occur between these new levels (for detail see e.g. [37,38]). The splitting of the ${}^4I_{13/2}$ and ${}^4I_{15/2}$ levels is governed by the local electric field strength and thus, the intensity and position of the single peaks in the ${}^4I_{13/2} \rightarrow {}^4I_{15/2}$ band changes from host to host (from surrounding to surrounding). This trend is observed in figure 5 upon the annealing temperature as well. The ${}^4I_{13/2} \rightarrow {}^4I_{15/2}$ band exhibits no peaks after the drying at 120 °C. Even the annealing at 250 °C has effect of the very small increase in intensity. The air annealing at 350 °C causes very strong raise of the band magnitude and the A1-A4 peaks become observed. At 400 °C, the intensity of the band reaches maximum. Annealing at 500 °C causes drop of the intensity down to the level of the 250 °C, however, some peak structure still can be detected. Further annealing at 600-700 °C results in monotonous increase in intensity of the new B1-B8 peaks. The B1-B8 peaks are narrower than the A1-A4 ones and their spectral position is different. Since the annealing was performed in air (oxidizing atmosphere), the $Er^{3+} \rightarrow Er^{2+}$ charge state transformation can hardly be expected. But $Er^{2+} \rightarrow Er^{3+}$ and $Er^{3+} \rightarrow Er^{4+}$ is, in principle, possible

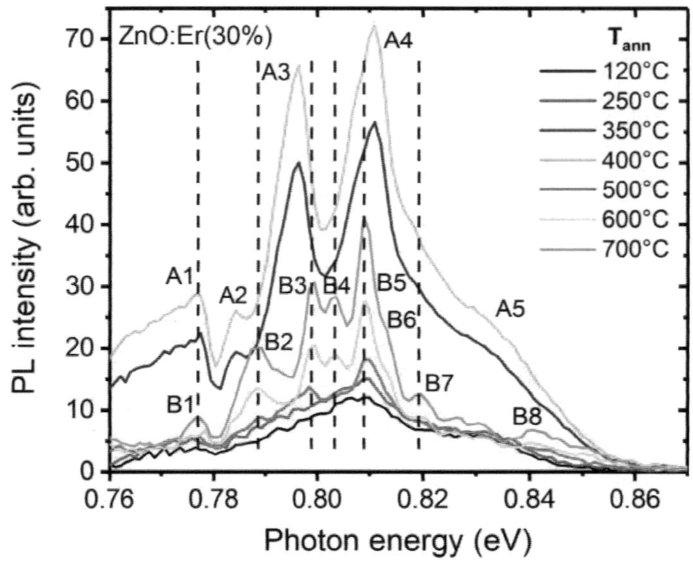

Figure 5. Er^{3+} emission, the $^4I_{13/2} \rightarrow {}^4I_{15/2}$ transition, measured at room temperature. It is split due to numerous transitions between the Stark levels [37,38]. Dashed vertical lines serve to indicate the shift of the peaks upon the annealing temperature T_{ann}. A1-A5 and B1-B8 indicate peaks for T_{ann} within 350-400 °C and 500-700 °C, respectively.

as discussed e.g. for nanocrystalline CeO_2:Er [39]. Probably, the $Er^{3+} \rightarrow Er^{4+}$ transformation happens at the step of the Er^{3+} emission decrease and two different $Er^{2+} \rightarrow Er^{3+}$ transformations occur at the steps of the Er^{3+} emission increase. Altogether this may point out to the existence of one site for Er^{3+} and two different sites for Er^{2+} ions (two rising tendencies in figure 5 related to the change in the local surrounding of the Er^{3+} ions).

3.4. PL and RL decay kinetics of the E2 and E3 bands

To study the kinetics of the E2 (1.82 eV) and E3 (3.25 eV) emission bands (see figure 4), the PL (excited with 310 nm laser light) and RL (excited with X-ray) decay curves were measured, respectively. The corresponding curves were obtained in the ZnO:Mo(10%) and ZnO:Er(30%) annealed at 350 °C, where the E2 and E3 bands were possible to measure, especially taking into account that the E3 band reaches maximum upon the annealing at this temperature. The curves were fitted and are demonstrated in figure 6. The decay of the E3 band was fitted with three exponentials in both ZnO:Mo(10%) and ZnO:Er(30%) (figures 6(a),(b)).

The decay curves in figure 6 have been fitted by using the following expression (convoluted with the instrumental response function)

$$I(t) = \sum_{i=1}^{n} A_i \exp\left(-\frac{t}{\tau_i}\right) + I_0, \tag{1}$$

where $I(t)$, A_i, and τ_i are the intensity varying in time (the number of photons detected in the given time interval), the amplitude of a single component and the corresponding decay time. n is the number of components. In the case of the curves in figures 6(a),(b) is $n = 3$. The characteristic decay times were found as follows: (i) $\tau_1 = 49$ ps, $\tau_2 = 571$ ps, and $\tau_3 = 16.4$ ns in ZnO:Mo(10%) (figure 6(a)); (ii) $\tau_1 = 53$ ps, $\tau_2 = 615$ ps, and $\tau_3 = 18.9$ ns in ZnO:Er(30%) (figure 6(b)). Apart from the very fast ps components, the 16.4 ns and 18.9 ns components are, most probably, the contributions from some defect with emission overlapping with the E3 band in both materials. The 571 ps determined in the

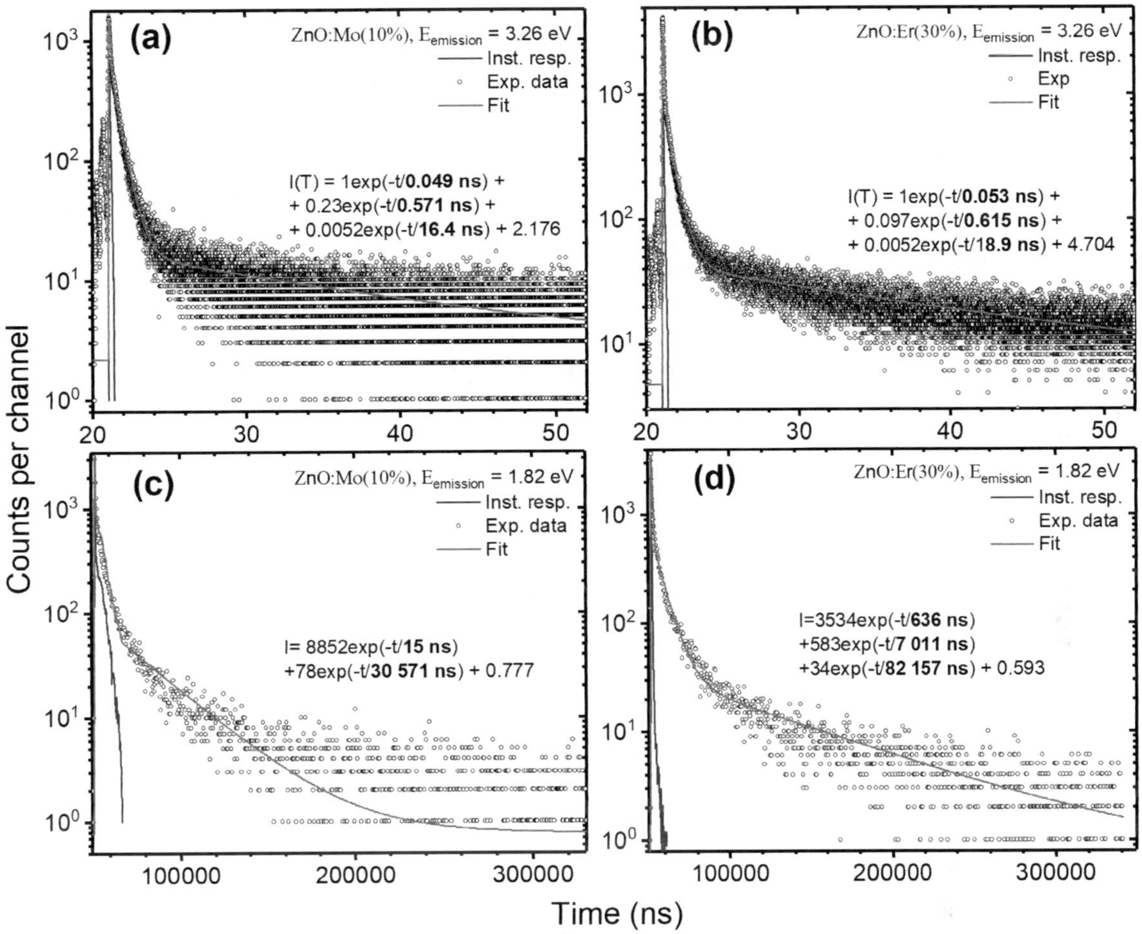

Figure 6. Room temperature decay curves measured with the X-ray excitation in (a) ZnO:Mo(10%) and (b) ZnO:Er(30%) at the E3 band (E_{emission} = 3.26 eV, $\lambda_{\text{emission}}$ = 380 nm). Room temperature decay curves measured with the 310 nm excitation at the E2 band (E_{emission} = 1.82 eV, $\lambda_{\text{emission}}$ = 680 nm) in (c) ZnO:Mo(10%) and (d) ZnO:Er(30%).

ZnO:Mo(10%) and 615 ps determined in the ZnO:Er(30%) can be considered as the real decay constants of the E3 exciton-related bands. So, one may expect the influence of the dopants on the exciton decay kinetics rate. These values are of the same order of magnitude as reported in literature for the exciton-related emission in ZnO [36].

The decay of the E2 band is double exponential in the ZnO:Mo(10%) (figure 6(c)) whereas it is three exponential in the ZnO:Er(30%) (figure 6(d)). They were fitted by using equation (1) with $n = 2$ and $n = 3$, respectively. The characteristic decay times were determined as follows: (i) τ_1 = 15 ns, τ_2 = 30.6 μs in ZnO:Mo(10%) (figure 6(c)); (ii) τ_1 = 636 ns, τ_2 = 7 μs, and τ_3 = 82.2 μs in ZnO:Er(30%) (figure 6(d)). These decay times are mostly of the submicrosecond or microsecond order. The largest decay time τ_2 determined in the ZnO:Mo(10%) is about 2.7 times smaller than that (τ_3) in the ZnO:Er(30%) whereas, the middle component of τ_2 is approximately 4 times smaller in the ZnO:Er(30%) as compared to the τ_1 in ZnO:Mo(10%). The smallest decay time τ_1 determined in the ZnO:Mo(10%) is by more than one order of magnitude smaller than the smallest decay time τ_1 in the ZnO:Er(30%). Due to the existence of two or three exponential components in the decay curves (see figures 6(c),(d)), which presumably originate from the similar emission mechanisms, the mean PL decay times were assessed by using the following expression

$$\tau_{mean} = \left(\sum_{i=1}^{n} A_i \tau_i \right)^{-1} \sum_{i=1}^{n} A_i \tau_i^2 \, . \tag{2}$$

They are almost the same in both samples: τ_{mean}(ZnO:Mo(10%)) = 28.96 μs and τ_{mean}(ZnO:Er(30%)) = 28.34 μs. All of this may benefit to the suggestion of strongly overlapping multicomponent structure of the E2 band where each component is differently sensitive to annealing and dopants. Note, that the E2 bands demonstrated different dependence on the annealing temperatures in the ZnO:Mo(10%) and ZnO:Er(30%) (figures 4(d),(f)).

3.5. Peculiarities of radioluminescence

RL can afford the general image of the processes occurring in the samples and therefore, the luminescence spectra in ZnO:Mo(10%)/(30%) and ZnO:Er(30%) was also measured under the X-ray excitation. The spectra are shown in figure 7.

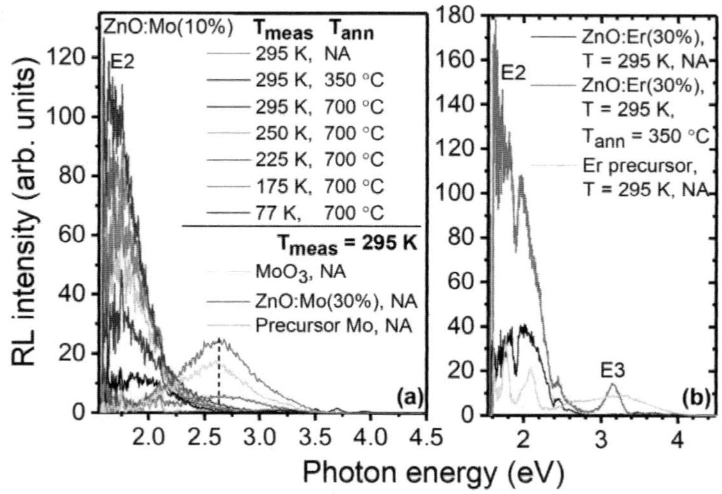

Figure 7. (a) RL spectra measured at different temperatures (T_{meas}) after annealing in air at chosen temperatures (T_{ann}) in ZnO:Mo(10%). RL spectra of ZnO:Mo(30%), MoO_3, and Precursor Mo samples were measured at T_{meas} = 295 K. Dashed vertical line emphasizes the same emission in ZnO:Mo(30%) and Precursor Mo. (b) RL spectra of ZnO:Er(30%) sample, not annealed and after annealing in air at 350 °C. RL spectrum of Precursor Er is also plotted. The E2 and E3 bands are indicated. NA stands for "not annealed".

RL spectrum was also measured in MoO_3 nanorods (see SEM image in figure 8). It is shown in figure 7(a) for comparison as well. The RL signal from MoO_3 nanorods is negligible as compared to the strength of the emission measured in the ZnO:Mo(10%)/(30%) samples and therefore, the creation of the MoO_3 in the ZnO:Mo(10%)/(30%) was excluded. By comparing the RL observed in the as-grown ZnO:Mo(10%)/(30%) samples and Precursor Mo, one can see, that they have identical spectral position, shape, and width. Therefore, the observed emission in the as-grown ZnO:Mo(10%)/(30%) samples was attributed to the Precursor Mo remained in the materials. The RL intensity in the as-grown ZnO:Mo(10%) is approximately three times lower than the RL intensity in the as-grown ZnO:Mo(30%) indicating larger amount of the Precursor Mo remained in the latter. This correlates very well to the observations done analysing the corresponding XRD patterns (figure 2(a)) and Raman spectra (figure 2(b)). Annealing in air at 350 °C annihilates the Precursor Mo emission in the ZnO:Mo(10%) sample and the E2 band observed also in PL can now be observed. This makes

Figure 8. SEM image of the MoO_3 sample.

evidence for the diminishing/decay of the Precursor Mo caused by the interaction with air. Further annealing at 700 °C improves the E2 band intensity approximately two times as expected from PL measurements (figure 4(d)). In order to study the behaviour of this band upon the temperature of observation, the sample was cooled down to 77 K stepwise and thus the temperature dependence of the E2 band was obtained by measuring RL spectrum at every step. The corresponding spectra are shown in figure 7(a) for chosen temperature points. They demonstrate monotonous increase in the intensity revealing negligible changes in the band position and shape. This allows to suggest that at least down to 77 K no extra charge or energy transfer appear to affect the source(s) of the band.

Since RL of the Precursor Mo (figure 7(a)) was very strong in the ZnO:Mo(30%) and the ZnO phase was hardly detected in this sample according to XRD and Raman spectroscopy (see figure 2), the annealing procedures were not carried out.

The RL spectrum of the as-grown ZnO:Er(30%) seems to consists of one band, E1, with the maximum at 1.91 eV that is split in two with a dip having minimum at 1.9 eV (see figure 7(b)). The RL spectrum measured in the Precursor Er is composed of two narrow bands and one broad band (2.2-4 eV) peaking at 1.74 eV, 2.05 eV, and 3.25 eV (figure 7(b)). By comparing the obtained RL spectra the observed dip at 1.9 eV was thus attributed to the mixing of the E1 band with the bands from Precursor Er. However, absence of the broad band (2.2-4 eV) with the maximum at 3.25 eV in the RL spectrum of the ZnO:Er(30%) may point out to the somehow modified Precursor Er remained in the material. Maybe it is an additional phase indicated by the XRD and Raman spectroscopy studies (figure 2) created from the Precursor Er, probably, by adsorbing/desorbing additional H_2O molecule to/from the original formulae (see subsection 2.1) in the solution during the hydrothermal growth.

3.6. Characterization of paramagnetic states by EPR

EPR technique was used in order to find out which paramagnetic particles are present in ZnO:Mo(10%), ZnO:Mo(30%), and ZnO:Er(30%) microrods and MoO_3 nanorods (for comparison) and possibly to determine correlations with optical measurements above. The EPR spectra measured in the ZnO:Mo(10%) are shown in figure 9. Prior to air annealing only EPR signal at g factor $g = 1.954$ was detected. It is typical resonance for ZnO reported in many other works [15-18,21]. Its g factor value varies from 1.95 to 1.96 in different ZnO materials [36]. Its origin was deduced to be F^+ centre [19,20], an electron captured by an oxygen vacancy. The signal is getting a bit stronger after annealing in air at T_{ann} = 250 °C with consequent intensity drop at T_{ann} = 350-450 °C. The rise of the spectrum intensity is observed after further annealing at T_{ann} = 500-600 °C. The increase of the signal at T_{ann} = 250 °C can, in principle, be explained as follows. The material can have some number of empty oxygen vacancies and imperfections which can change their charge state in the oxidizing atmosphere. Some part of the imperfections may supply the vacancies with electrons (one electron per one vacancy) during the annealing and thus the new F^+ centres of the same sort are created contributing to the increase of the F^+ EPR signal intensity. Annealing at higher temperatures (T_{ann} = 350-450 °C) causes partial refilling of the vacancies with oxygen from air thus annihilating the F^+ centres and the F^+ signal is getting significantly weaker (figure 9(a)). This tendency was observed and reported previously in the undoped ZnO nanorods [21]. However, some small number of F^+ centres, most probably, from the microrods bulk, withstand the annealing at even higher T_{ann} = 500-600 °C as long as the F^+ signal is slightly rising again indicating the creation of new F^+ centres of the same sort. The material may contain some amount of F centres, two electrons trapped in one vacancy. They are not paramagnetic and thus cannot be detected by EPR. Probably, by annealing at T_{ann} = 500-600 °C, some part of these centres loses electrons (one from each centre) and thus the F^+ centres are created contributing to the increase of the F^+ centres signal intensity. No correlation can thus be found with PL (see subsection 3.5 and figure 4(d)).

Besides the described above transformations another change occurs in the ZnO:Mo(10%) EPR spectrum under annealing: a new signal appears at T_{ann} = 350 °C (figures 9(a),(b)). This signal is centred at g factor g = 1.905 (3520 G). Remarkably, it has never been observed before in the undoped ZnO upon air annealing [21]. Therefore, it was related to the Mo ion. Since no other lines of this sort were observed in EPR spectra and the position of the g = 1.905 resonance line was unchanged upon

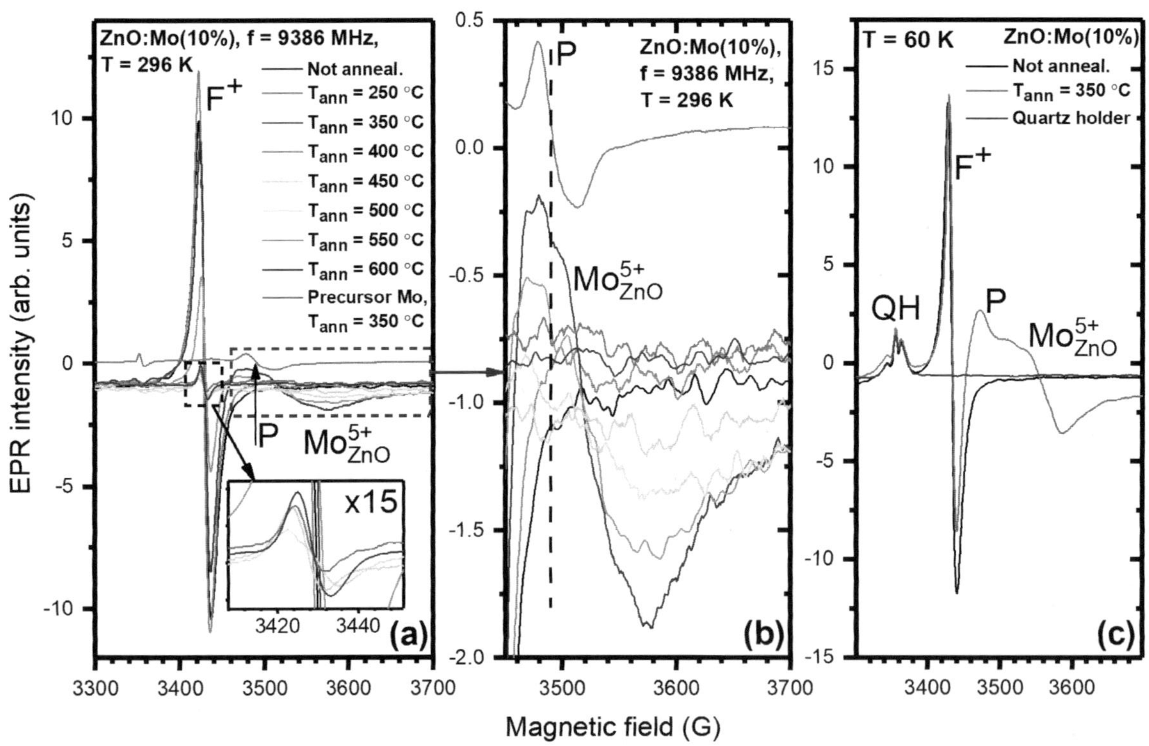

Figure 9. (a) EPR spectra measured at T = 296 K in ZnO:Mo(10%) before and after annealing in air (the annealing temperatures T_{ann} are given in a legend). The spectrum measured in Precursor Mo is also plotted. Blue dashed region, zoomed in the panel (b), indicates the signal possibly originating from Mo^{5+} ion incorporated (deposited on surface) into ZnO microrods bulk (Mo^{5+}_{ZnO}). (c) EPR spectra measured at T = 60 K in ZnO:Mo(10%) before and after annealing in air at T_{ann} = 350 °C. The spectrum measured in an empty quartz holder is also plotted. P and QH indicate the signals coming from Precursor Mo and quartz holder, respectively.

IOP Conf. Series: Materials Science and Engineering 1050 (2021) 012002 doi:10.1088/1757-899X/1050/1/012002

the sample cooling down to 10 K, the Mo charge state is expected to be "5+". Moreover, the *g* factor has the value typical for Mo^{5+} measured in different oxide materials [40-44]. The $g = 1.905$ signal is broad and obviously multicomponent having rather powder-like shape [45]. To increase the signal intensity the ZnO:Mo(10%) sample was measured also at $T = 60$ K (figure 9(c)). All this allowed to suppose the Mo^{5+} to be embedded into some host. In order to find out whether the signal is coming from microrods or Precursor Mo, the latter underwent annealing in air at $T_{ann} = 350$ °C and then measured. Corresponding EPR spectrum is demonstrated in figures 9(a),(b) as well.

It contains also one signal at *g* factor $g = 1.921$ (3491 G), indicated P in figures 9(a),(b), presumably originating from Mo^{5+} as well. This *g* factor is larger (1.921 > 1.905) and the signal shape is different than that of the $g = 1.905$ signal. Therefore, the $g = 1.905$ signal cannot originate from the Precursor Mo. However, it is known that Precursor Mo is unstable decomposing under heating on $MoO_{2.5}(OH)_{0.5}$ (annealing at ca. 325 °C to 475 °C), $MoO_{2.69}(OH)_{0.3}$ (annealing at ca. 575 °C) and MoO_{3-x} (annealing at ca. 675 °C) [29]. Therefore, the MoO_3 nanorods were also measured. Corresponding EPR spectra are shown in figure 10 for different temperatures.

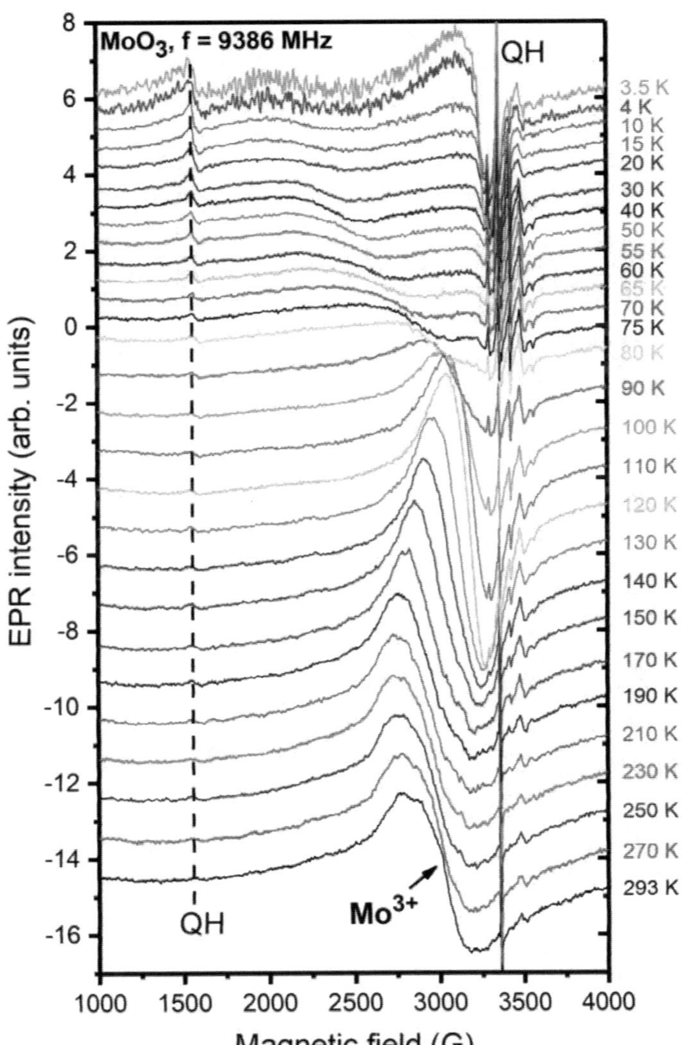

Figure 10. EPR spectra measured at different temperatures (293 - 3.5 K) in MoO_3 nanorods. "QH" stands for "quartz holder".

There the dominating resonance at $g = 2.21$ at 296 K, most probably, belongs to Mo^{3+} [46] since it is too large for the Mo^{5+} [40-44]. Moreover, it exhibits temperature dependence typical to some extent for magnetic ordering (the line is getting broader and shifted upon the cooling) [47]. Nothing similar was observed in the case of the ZnO:Mo(10%).

All this leads to conclusion that the $g = 1.905$ EPR signal belongs to the Mo^{5+} from the microrods (Mo^{5+}_{ZnO}). Further annealing at T_{ann} = 400-600 °C results in the Mo^{5+}_{ZnO} signal decrease and complete disappearance. This allows to suppose the starting charge state (before the treatment) of the Mo to be either "2+" ($4d^4$ outer shell) or "4+" ($4d^2$), both were not detected in the X-band EPR because of large spin-orbit coupling constant (SOC), 695 and 950 cm^{-1}, respectively. Annealing in air causes stepwise oxidation to the highest possible "6+" charge state, which is not paramagnetic and thus, the Mo^{5+}_{ZnO} signal disappears. No traces of Mo^{3+} were detected during the annealing procedure. This benefits to the starting Mo^{4+} charge state in the ZnO before the treatment. Remarkably, the Mo^{4+} ion has ionic radius 0.65 Å in the sixfold ligand coordination [48] and thus it is expected to be even smaller in the fourfold one (unfortunately, no data on this subject are available in literature) in ZnO. The Zn^{2+} ionic radius is 0.6 Å in the fourfold coordination whereas it is 0.74 Å in the sixfold one. Thus, it seems possible for the Mo^{4+} to be embedded into the ZnO host at Zn site.

EPR spectra measured in the ZnO:Mo(30%) sample before and after annealing in air at different temperatures are demonstrated in figure 11. The spectrum measured in the ZnO:Mo(30%) nanorods was marked as Mn^{2+}_1 in figure 11(a). Larger number of transitions (better resolved structure) becomes visible after cooling down to T = 60 K. They were most easily identified due to the forbidden transition (F.T.) in figure 11(a). It is situated at the half-field position (full-field position corresponds

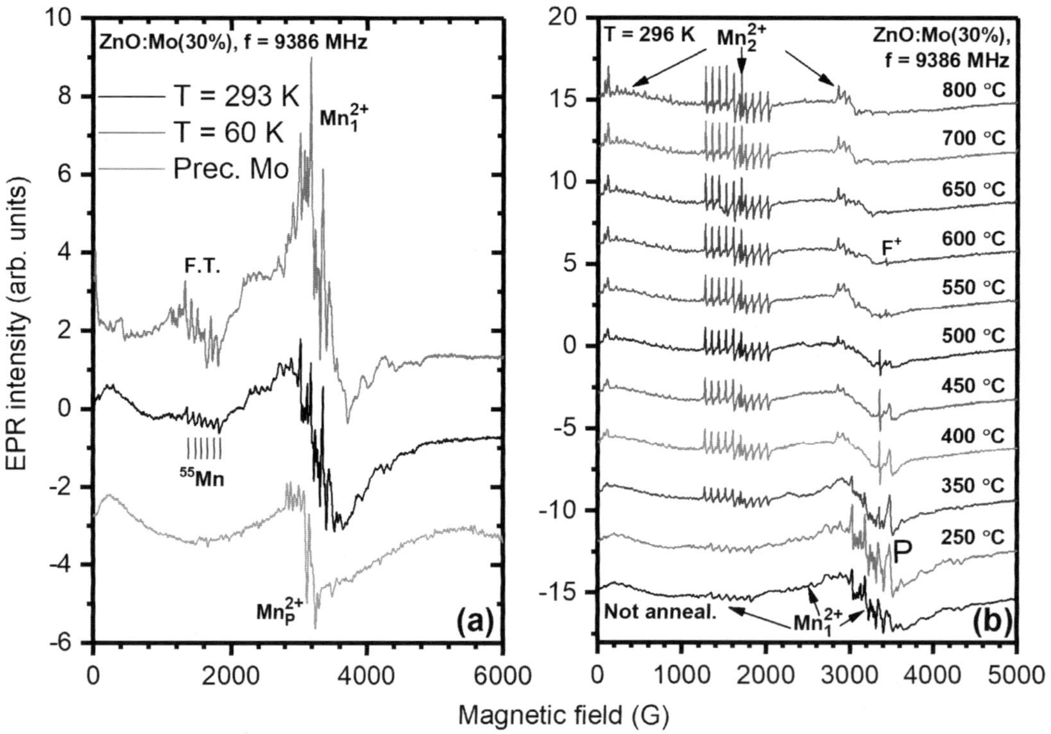

Figure 11. (a) EPR spectra measured at T = 60 and 296 K in ZnO:Mo(30%) before annealing in air. P and F.T. indicate the signals coming from Precursor Mo and forbidden transition of Mn^{2+}. The sextet of lines due to ^{55}Mn nucleus is indicated by the set of vertical lines. Mn^{2+} EPR spectrum measured in Precursor Mo is designated as Mn^{2+}_P. (b) EPR spectra measured at T = 296 K in ZnO:Mo(30%) before and after annealing in air (the annealing temperatures are given in a legend).

to the *g* factor about 2 typical for the Mn^{2+}, the *S* ion with $^6S_{5/2}$ ground level, i.e., quenched orbital moment [44,45]). F.T. had the most pronounced structure, i.e. hyperfine structure (HFS) due to ^{55}Mn nucleus with nuclear spin $I = 5/2$ and 100% abundance (figure 11(a)) as compared to the full-field position (1/2 ↔ -1/2 electron spin transition) which is too intermixed with other transitions either allowed (in general there should be 5 of them [44,45]) or forbidden. The splitting between lines in the F.T. sextet was about 95 G, typical value for the Mn^{2+} measured in other materials [44]. Annealing in air at elevated temperatures results in transformation of the Mn_1^{2+} signal shown in figure 11(a) to different Mn_2^{2+} spectrum (figure 11(b)). The new spectrum was identified due to the two pronounced sextets of lines (with about 95-100 G spacing) centred at approximately 1600 G magnetic field (figure 11(b)) using the same considerations as above for the Mn_1^{2+} spectrum identification. Manganese is an uncontrolled impurity, most probably, originating from Precursor Mo since the typical Mn^{2+} EPR spectrum [49] was measured there as well (for more details see figure 11(a) where it is designated as Mn_P^{2+}). The Mn^{2+} EPR spectrum has never been detected in the rest of samples.

No sign of the typical signal of F^+ centre was detected before annealing. It becomes visible at *g* factor $g = 1.956$, but negligibly weak, after the annealing at 500 °C with the intensity approximately doubled for 600 °C. Note, that the g factor is a bit shifted in the ZnO:Mo(30%) as compared to the ZnO:Mo(10%) above. Above the annealing temperature of 600 °C this signal is vanished (see figure 11(b)). Moreover, the P spectrum from Precursor Mo appeared after annealing at 250 °C and disappeared completely at 550 °C. All this finds correlate with the low content of ZnO phase predicted by XRD and Raman analyses (see figure 2 in subsection 3.1).

Oppositely, EPR spectra measured in the ZnO:Er(30%) sample exhibit strong signal from the F^+ centre (see figure 12(a)) before annealing (*g* factor $g = 1.953$). The subsequent annealing in air at

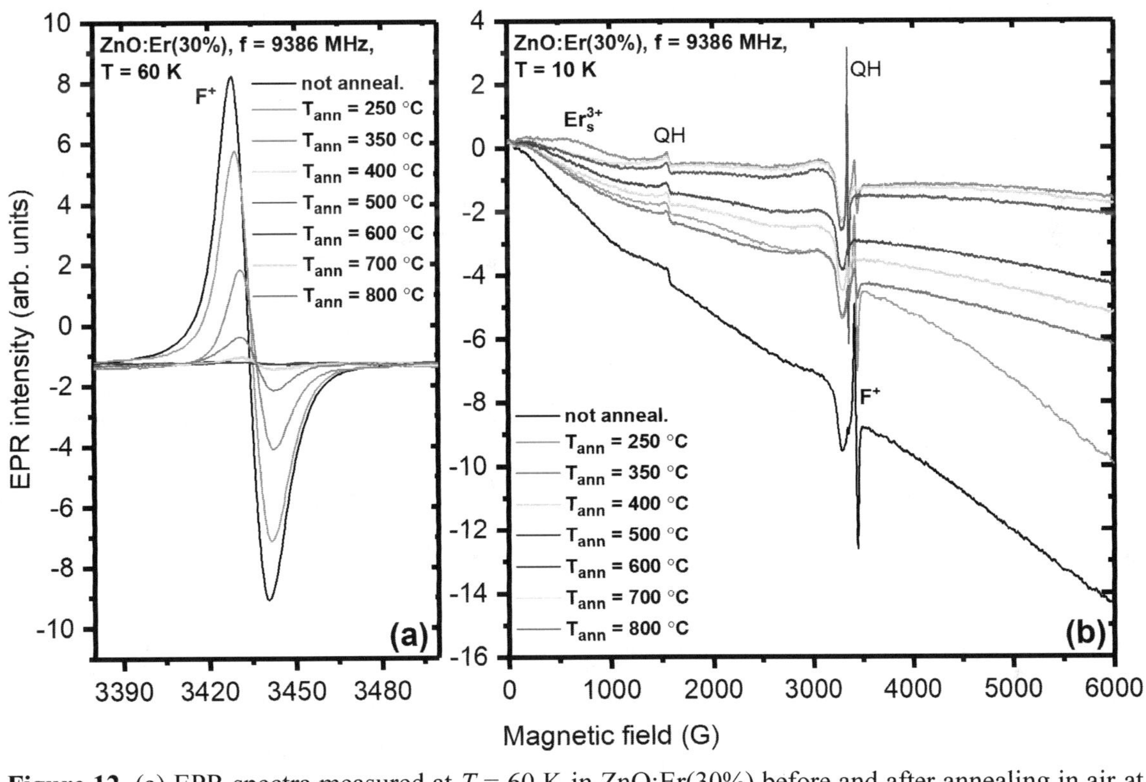

Figure 12. (a) EPR spectra measured at $T = 60$ K in ZnO:Er(30%) before and after annealing in air at temperature T_{ann}. (b) EPR spectra measured at $T = 10$ K in ZnO:Er(30%) before and after annealing in air at temperature T_{ann}. QH indicates the signals coming from quartz holder. Er_S^{3+} indicates the signal from Er^{3+} on the surface of ZnO microrods.

elevated temperatures causes drop of the intensity and almost total disappearance of the signal within the 250-600 °C range of annealing temperatures. Above 600 °C the signal starts to rise again.

To measure rare-earth Er^{3+} EPR spectra, the sample should be cooled down to 10 K since shielded 4f shell in this case results in very short time of spin-lattice relaxation making the spectra very broad at room temperature [45]. The spectra measured at 10 K are shown in figure 12(b) for the same set of annealing temperatures as in figure 12(a). Prior to the annealing the spectrum is very broad (extends the magnetic field range allowed by the magnet of spectrometer) and strong. It was attributed to Er^{3+} since the correlation between the doping level and EPR intensity was observed. Moreover, according to the absorption spectrum in figure 3 Er^{3+} is present in the sample. It is also not excluded that the observed EPR spectra shown in figure 12(b) are the combination of contributions from Er^{2+}, Er^{3+}, and Er^{4+}. Annealing makes the signal even broader while the peak-to-peak intensity goes down after the annealing within the 250-400 °C.

Linewidth of the EPR signal is proportional to the paramagnetic particles' concentration [49]. Therefore, the broader signal the larger number of paramagnetic particles. This behaviour correlates well to the observed tendencies in the Er^{3+} emission (figure 5). Therefore, similarly one may expect the $Er^{3+} \rightarrow Er^{4+}$ transformation at the step of the Er EPR signal narrowing and two different $Er^{2+} \rightarrow Er^{3+}$ transformations at the steps of the Er EPR signal broadening. This supports the supposition of the existence of one site for Er^{3+}, probably Er_S^{3+} (figure 12(b)), and two different sites for Er^{2+} ions. Basing on the analysis done for the infrared Er^{3+} emission (figure 5) and Er EPR spectra (figure 12(b)) it may be concluded that one of the Er^{2+} ions could be implanted inside the ZnO host. Er^{2+} and Zn^{2+} both have the same charge state.

4. Conclusions

According to SEM, the ZnO nanorods when alloyed with Mo or Er become microrods. EPR confirmed the existence of F^+ centres common for ZnO-based structures in the ZnO:Er(30%) microrods whereas in the ZnO:Mo(30%) this kind of defect was almost absent and appeared only after annealing in air. Annealing in air at elevated temperatures results in the reduction of F^+ centres in the ZnO:Mo(10%) and ZnO:Er(30%). Moreover, Er^{3+} EPR signal also undergoes changes under annealing getting broader at 250-400 °C and then narrower again at 500-800 °C. This allowed to suggest oxidation of Er ions in the ZnO:Er(30%) sample in correlation with similar behaviour of infrared emission of Er^{3+}. The existence of Er^{2+} ions inside the ZnO host could be concluded. Annealing also results in the Mo^{5+} appearance presumably in the ZnO:Mo(10%) microrods bulk. Red luminescence peaking near 680 nm appears under air annealing (to the best of our knowledge it has not been observed before) in all studied samples regardless the dopant origin and doping level. However, its origin remains unknown. The exciton-related band at 380 nm never observed in the samples before the annealing appears under the annealing in air reaching maxima at 350 °C in ZnO:Mo(10%) and ZnO:Er(30%). The decay time of this exciton-related emission measured in both samples was about 600 ps. No such band was observed in the ZnO:Mo(30%) sample at all.

Acknowledgements

This work was supported by the Czech Science Foundation project No. 20-05497Y, the Ministry of Education, Youth and Sports of Czech Republic under the project SOLID21 No. CZ.02.1.01/0.0/0.0/16_019/0000760, and by the Scientific Grant Agency of Ministry of Education, Science, Research and Sport of Slovak Republic and Slovak Academy of Sciences (VEGA 2/0157/20).

References

[1] 2017 *Nanocomposite, Ceramic and Thin Film Scintillators* ed M Nikl (Singapore: Pan Stanford Publishing Pte. Ltd)
[2] Wilkinson J, Ucer K B and Williams R T 2004 *Radiat. Meas.* **38** 501–5
[3] Abrahams S C and Bernstein J L 1969 *Acta Cryst.* B **25** 1233-6

[4] Burešová H *et al* 2016 *Optics Express* **24** 15289-98
[5] Shi Q, Wang Ch, Li Sh, Wang Q, Zhang B, Wang W, Zhang J and Zhu H 2014 *Nanoscale Res. Lett.* **9** 480
[6] Ningthoujam R S 2008 *J. Nanosci. Nanotech.* **8** 3059–62
[7] Savoyant A, Alnoor H, Bertaina S, Nur O and Willander M 2017 *Nanotech.* **28** 035705
[8] Reddy A J, Kokila M K, Nagabhushana H, Rao J L, Nagabhushana B M, Shivakumara C and Chakradhar R P S 2011 *Spectrochimica Acta Part* A **79** 476–80
[9] Orlinskii S B, Schmidt J, Baranov P G, Hofmann D M, De Mello Donegá C and Meijerink A 2004 *Phys. Rev. Lett.* **92** 047603
[10] Baranov P G, Romanov N G, Bundakova A P, Orlinskii S B, De Mello Donegá C and Schmidt J 2016 *Mater. Today: Proc.* **3** 816–24
[11] Ishizumi A, Taguchi A, Yamamoto A and Kanemitsu A 2005 *Thin Solid Films* **486** 50–2
[12] Lo J-W, Lien W-Ch, Lin Ch-An and He Jr-H 2011 *Appl. Mater. Interfaces* **3** 1009–14
[13] Babayevska N, Peplinska B, Jarek M, Yate L, Tadyszak K, Gapinski J, Iatsunskyia I and Jurga S 2016 *RSC Adv.* **6** 89305.
[14] Wei Xi, Wang W and Chen Ke 2013 *J. Phys. Chem.* C **117** 23716–29
[15] Vorobyeva N, Rumyantseva M, Filatova D, Konstantinova E, Grishina D, Abakumov A, Turner S and Gaskov A 2013 *Sensors and Actuators* B **182** 555–64
[16] Vorobyeva N, Rumyantseva M, Konstantinova E, Grishina D and Gaskov A 2011 *Proc. Eng.* **25** 296–9
[17] Neykova N, Chang Yu-Ying, Buryi M, Davydova M, Kucerkova R, Simek D, Remes Z and Pop-Georgievski O 2019 *Appl. Surface Sci.* **472** 105–11
[18] Hofmann D M, Hofstaetter A, Leiter F, Zhou H, Henecker F, Meyer Bruno K I, Orlinskii S B, Schmidt J and Baranov P G 2002 *Phys. Rev. Lett.* **88** 045504
[19] 1968 *Physics of color centers* ed W B Fowler (New York and London: Academic press)
[20] Laguta V, Buryi M, Arhipov P, Sidletskiy O, Laguta O, Brik M G and Nikl M 2020 *Phys. Rev.* B **101** 024106
[21] Buryi M, Babin V, Chang Y-Y, Remeš Z, Mičová J and Šimek D 2020 *Appl. Surface Sci.* **525** 146448
[22] Bai Sh, Chen Ch, Tian Ye, Chen S, Luo R, Li D, Chen A and Liu Ch Ch 2015 *Mater. Res. Bulletin* **64** 252–6
[23] Mooney J and Kambhampati P 2013 *J. Phys. Chem. Lett.* **4** 3316–8
[24] Chang Y-Y, Remes Z and Micova J 2020 *NANOCON 2019 - Conference Proceedings* 11th International Conference on Nanomaterials - Research and Application 16-18 October 2019 Hotel Voronez I, Brno, Czech Republic (Ostrava: Tanger Ltd.) pp 221–5
[25] Remes Z, Micova J, Krist P, Chvatil D, Effenberg R and Nesladek M 2015 *Phys. status solidi* A **212** 2519–24
[26] Remes Z, Babchenko O, Varga M, Stuchlik J, Jirasek V, Prajzler V, Nekvindova P and Kromka A 2016 *Thin Solid Films* **618** 130–3
[27] Mičová J, Buryi M, Šimek D, Drahokoupil J, Neykova N, Chang Y-Y, Remeš Z, Pop-Georgievski O, Svoboda J and Im Ch 2018 *Appl. Surface Sci.* **461** 190–5
[28] Marsh J K 1941 *J. Chem. Soc.* 561-2
[29] Chithambararaj A, Bhagya Mathi D, Rajeswari Yogamalar N and Chandra Bose A 2015 *Mater. Res. Express* **2** 055004
[30] Pimentel A, Nunes D, Duarte P, Rodrigues J, Costa F M, Monteiro T, Martins R and Fortunato E 2014 *J. Phys. Chem.* C **118** 14629–39
[31] Murugan R and Chang H 2001 *J. Chem. Soc., Dalton Trans.* 3125–32
[32] Trejo Aguilar I E 2012 *Crecimiento de películas de ZnSe umpurificadas con tierras raras y studio de sus propiedades opticas y eléctricas* (Tuxtla Gutiérrez: Master thesis) p 48
[33] Ren T, Baker H R and Poduska K M 2007 *Thin Solid Films* **515** 7976–83

[34] Wei T, Tian Y, Chen F, Cai M, Zhang J, Jing X, Wang F, Zhang Q and Xu Sh 2014 *Sci. reports* **4** 6060
[35] Roth A P, Webb J B and Williams D F 1981 *Solid State Comm.* **39** 1269–71
[36] Klingshirn C F 2010 *Zinc Oxide: from Fundamental Properties towards Novel Applications* (London: Springer)
[37] Michel J, Benton J L, Ferrante R F, Jacobson D C, Eaglesham D J, Fitzgerald E A, Xie Y-H, Poate J M and Kimerling L C 1991 *J Appl. Phys.* **70** 2672
[38] Venkatramu V *et al* 2012 *J. Mater. Chem.* **22** 13788
[39] Rakhmatullin R M, Aminov L K, Kurkin I N and Pöppl A 2015 *Appl. Magn. Reson.* **46** 741–48
[40] Buryi M, Laguta V, Fasoli M, Moretti F, Trubitsyn M, Volnianskii M, Vedda A and Nikl M 2017 *J. Lumin.* **192** 767–74
[41] Buryi M *et al* 2019 *J. Lumin.* **205** 457–66
[42] Spassky D A *et al* 2013 *Opt. Mater.* **35** 2465–72
[43] Buryi M, Spassky D A, Hybler J, Laguta V and Nikl M 2015 *Opt. Mater.* **47** 244–50
[44] 1999 *Handbook of Electron Spin Resonance* vol 2, ed C P Poole Jr and H A Farach (New York: Springer-Verlag)
[45] Abragam A and Bleaney B 1970 *Electron Paramagnetic Resonance of Transition Ions* vol 1 (Oxford: Clarendon Press)
[46] Gonzalez J R, Alcantara R, Tirado J L, Fielding A J and Dryfe R A W 2017 *Chem. Mater.* **29** 5886–95
[47] Díaz-Pardo R and Valenzuela R 2015 Characterization of magnetic phases in nanostructured ferrites by electron spin resonance *Advanced Electromagnetic Waves* ed S O Bashir (Rijeka: InTech) chapter 8 pp 210–37
[48] Shannon R D 1976 *Acta Crystallogr* A *Cryst. Phys., Diffr., Theor. Gen. Crystallogr.* **32** 751–67
[49] Altshuler S A and Kozyrev B M 1974 *Electron Paramagnetic Resonance in Compounds of Transition Elements* 2nd ed (New York: John Wiley & Sons, Inc.)

IOP Conf. Series: Materials Science and Engineering 1050 (2021) 012003 doi:10.1088/1757-899X/1050/1/012003

The design of an impedance tube and testing of sound absorption coefficient of selected materials

R Ďuriš and E Labašová

Slovak University of Technology in Bratislava, Faculty of Materials Science and Technology in Trnava, Ulica Jána Bottu 2781/25, 917 24 Trnava, Slovak Republic

rastislav.duris@stuba.sk

Abstract. The paper describes a design, building and testing a low-cost impedance tube for measuring the acoustic parameters of small testing samples. The measurement of sound absorption coefficient is based on transfer function and the two microphone method in accordance with ISO 10534-2 and ASTM E 1050 standards. With small configuration updates and after installation of additional microphones on measuring apparatus testing of complex acoustic material properties can be performed. The impedance tube has been used also as a teaching tool.

1. Introduction
The acoustic properties of materials such as sound absorption coefficient, reflection factor or transient loss (sound reduction) are determined experimentally. Several methods are in the role of determining the acoustic properties of materials on small samples. Procedure to ensure the measurement accuracy and repeatability of acoustic properties of materials, processing of measured results is prescribed by normative regulations in ISO 10534-1 (standing wave ratio method), ISO 10534-2, ASTM E 1050 (two microphone transfer function method for measurement of acoustic factor, sound absorption coefficient, specific acoustic impedance ratio, and specific acoustic admittance), and ASTM E 2611 (transient loss measurement) [1-3].

For testing the acoustic properties of materials according to the mentioned standards different impedance tubes were used. Most widely used is the set of Brüel&Kjær impedance tubes Type 4206 [4], which are something like industry standard in testing the acoustic material properties. Other commercially available tubes offer e.g. companies [5-8]. However, these commercially available tubes are expensive. Therefore, like other authors especially from universities, we went to develop a noncommercial tube. The design of low-cost impedance tubes suitable in the teaching process are presented in [9-10]. More complex solution is presented in works [11-13].

The design of a tube was carried out in accordance with ISO and ASTM Standards and also with respect to conclusions and recommendations resulting from the papers [14-15]. The procedure for measuring and processing of the results is described in [16-17].

2. Design of an impedance tube for the transfer function method
The impedance tube needs to be straight, with a smooth, nonporous inner wall. The material of the tube has to be strong and the thickness of the wall must prevent the vibrations within operating frequency range of the tube. The recommended value of the wall thickness is equal to 5 % of inner tube diameter. The design of the tube has to ensure that the upper limiting frequency of interest f_u is

lower than the lowest cut-off frequency of the tube. That condition allows to arise only plane propagating waves inside the tube. This state is expressed by [3] as:

$$d < K\frac{c}{f_u} \tag{1}$$

where *fu* is the upper working frequency of the tube, and c is the speed of sound in the air at actual temperature. For a circular cross-section of the tube, d [m] is its inner diameter. For the constant ASTM E 1050 [3] uses $K = 0.586$, whereas ISO 10534-2 [2] considers $K = 0.58$. For a rectangular cross-section, d is its largest section dimension and both standards use $K = 0.5$. The second condition to develop the plane wave in the tube is its sufficient length l. Minimum distance between sound source and first microphone (microphone A in Figure 1) must by equal to the inner diameter of the tube d, but it is recommended to meet the condition:

$$l - s - x_2 > 3d \tag{2}$$

Equations (1) and (2) are used to define the main dimensions of the impedance tube to avoid non-plane propagating waves, as required by measurements.

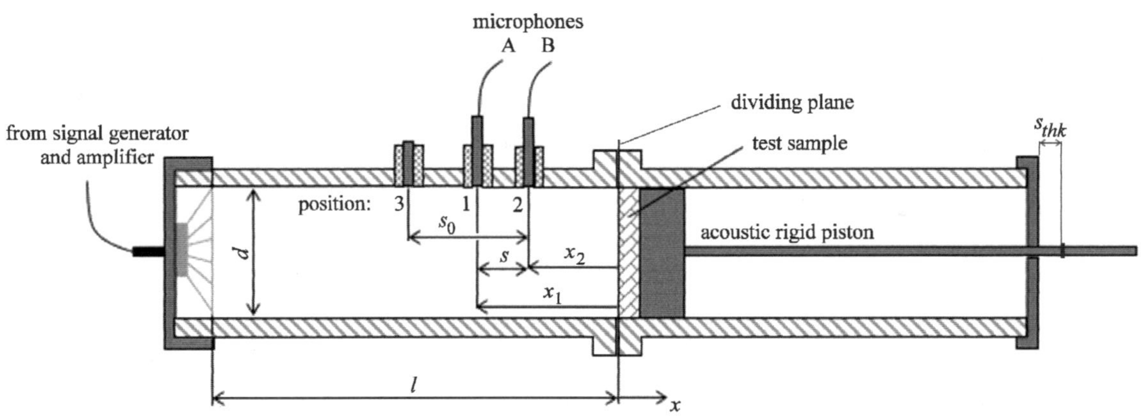

Figure 1. Design parameters of the impedance tube.

The both standards (ISO and ASTM) recommend using microphones of the same type designed for free sound field. Microphones membrane diameter d_{mic} must be much smaller (less than 20%) than the distance s between them, for a circular cross-section impedance tube to reduce the influence of their acoustic centers. Typically, 1/4 or 1/2 inch diameter size microphones are used for measurement. At the time of the tube design, there were the microphones B&K Type 4189-A-021 (1/2 inch microphone for open-field with pre-amplifier Type 2671, frequency range 20 Hz to 20 kHz, pre-polarized) available for measurement at our workplace. Spacing s between the measurement microphones in positions 1 and 2 governs the relations:

$$\begin{aligned} & d_{mic} < 0{,}2s & & \\ & f_u\, s < 0{,}45c & & \text{(ISO 10534-2)} \\ & f_u\, s < 0{,}40c & & \text{(ASTM E 1050).} \end{aligned} \tag{3}$$

Position of the third microphone with distance s_0 between point 2 and 3 (in Figure 1) depends on the lower limiting frequency f_l of the tube

$$s_0 > 0.05\frac{c}{f_l} \tag{4}$$

The upper f_u and lower f_l limit of the tube working frequency range can by calculated using relations:

$$f_u < K\frac{c}{d} \quad \text{and}$$

$$f_l > 0.01\frac{c}{s} \text{ (ASTM E 1050)} \quad \text{or} \quad f_l > 0.05\frac{c}{s} \text{ (ISO 10534-2).}$$

The spacing x_2 between the test sample and the closest microphone (microphone B in Figure 1) must prevent the proximity distortions in the acoustic field [14]:

- non-structured layer $x_2 > d/2$
- semi-lateral structured layer $x_2 > d$ (5)
- strongly asymetrical layer $x_2 > 2d$

Generally, the tube cross-section (diameter) must decrease if the maximum frequency of interest f_u increases. The non-plane waves always arise at the vicinity of the sound source and the test specimen surface. Therefore, the microphones cannot be placed close to the both tube ends. It is recommended to leave a distance of at least three tube diameters between the loudspeaker and its nearest microphone, and two tube diameters between the test specimen and the nearest microphone [2].

3. Transfer function method

The two-microphone impedance method is based on measuring the sound pressure in the impedance tube at two positions and calculation of the transfer function of acoustic pressure field between them. Figure 2 shows the apparatus for the measurement of acoustic properties of materials by a two-microphone method.

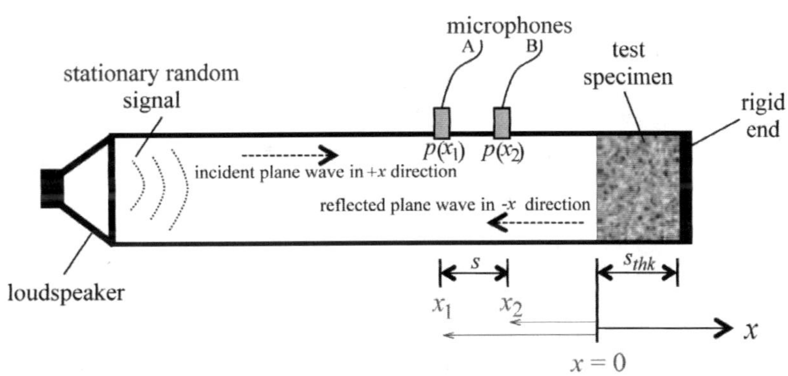

Figure 2. Scheme of two-microphone transfer function method.

The wave field in the tube is characterized by the acoustic pressure $p(x)$ of the incident wave $p_I(x)$ and the wave reflected from the sample $p_R(x)$. Total sound pressure $p(x)$ at any point in the tube is

$$p(x) = p_I(x) + p_R(x) = Ae^{-ikx} + Be^{ikx} \tag{6}$$

where A, B are amplitudes of the acoustic pressures of the incident and reflected wave and $\mathrm{i} = \sqrt{-1}$. Let $P(x,\omega)$ is the Fourier transform of $p(x)$ for the given location x, where ω is the angular frequency. The transfer function H_{12} of acoustic pressure field between positions 1 and 2 is defined by the complex ratio:

$$H_{12} = \frac{P(x_2,\omega)}{P(x_1,\omega)} = \frac{e^{-ikx_2} + re^{ikx_2}}{e^{-ikx_1} + re^{ikx_1}} \tag{7}$$

where r is the reflection factor of tested material and k is the wave number.

IOP Conf. Series: Materials Science and Engineering 1050 (2021) 012003 doi:10.1088/1757-899X/1050/1/012003

The reflection factor r is a complex number due to consideration of a phase shift of the waves. The estimation of the complex acoustic transfer function H_{12} can be carried out in three ways as follows:

$$H_{12} = \frac{S_{12}}{S_{11}} \quad \text{can be used generally,}$$

$$H_{12} = \frac{S_{22}}{S_{21}} \quad \text{recommended for cases where noise is at input e.g. if white noise is used to activate the acoustic waves in the tube, standard ASTM recommends to use this equation to determine the transfer function,} \tag{8}$$

$$H_{12} = \left(\frac{S_{12}}{S_{11}}\frac{S_{22}}{S_{21}}\right)^{1/2} \quad \text{recommended for cases involving noise is at input and output,}$$

where S_{12} is the cross-spectrum of the signal from microphones A and B, S_{11} is the autospectrum of the signal from microphone A, S_{22} is the autospectrum of the signal from microphone B, and S_{21} is the cross power spectral density of the signal from microphones B and A.

When using the two microphone technique, procedure of repeated measurements with interchanged channels for correcting the measured transfer function data must be used. When the test specimen is placed in the tube, two transfer functions H_{12}^{I} and H_{12}^{II} are estimated. For the first time, the microphones A, B are in positions 1, 2 (standard configuration I), as shown in Figure 1, and the transfer function H_{12}^{I} is determined. For the second measurement positions of microphones are switched on. Microphone A is in the position 2 and microphone B is in the position 1 (interchanged configuration II). Corrected transfer function H_{12} is computed using the equation:

$$H_{12} = \left(H_{12}^{\mathrm{I}}\, H_{12}^{\mathrm{II}}\right)^{1/2} = H_r + jH_i \tag{9}$$

where H_r is the real component and H_i is an imaginary part of the transfer function H_{12}.

The acoustic reflection factor r of the test specimen is from equation (7) defined as:

$$r = r(\omega) = \frac{H_{12} - e^{-jks}}{e^{jks} - H_{12}} e^{2jk(s+x_2)} = r_r + jr_i \tag{10}$$

The sound absorption coefficient α is the dimensionless quantity with the values in the range of $\langle 0, 1\rangle$, given as [2]:

$$\alpha = \alpha(\omega) = 1 - |r|^2 = 1 - r_r^2 - r_i^2 \tag{11}$$

Reflection factor r and sound absorption coefficient α vary with the frequency and angle of sound wave incidence. In the case, that the material perfectly reflects all incident sound waves from the surface, the sound absorption coefficient is $\alpha = 0$. In the case that the material absorbs all incident energy, the sound absorption coefficient is $\alpha = 1$. Therefore, the materials with porous or fibrous structure are recommended for the absorption of sound.

Specific acoustic impedance ratio Z is calculated from the equation:

$$Z = Z(\omega) = \rho c \frac{1+r}{1-r} \tag{12}$$

where ρ is the density of the air at actual temperature and the product (ρc) represents the characteristic impedance.

4. Building the impedance tube

The designed impedance tube is made of the thick walled HDPE pipe with the circular cross-section uniform along the length. The main design parameters of the tube are summarized in Table 1.

Table 1. Parameters of the impedance tube.

Parameter	Value
inner diameter of the tube d	0.088 m
tube wall thickness t	0.010 m
tube length l	0.758 m
distance x_2 of nearest microphone and sample[a]	0.100 m
distance s between microphone positions 2 and 1	0.065 m
distance s_0 between microphone positions 2 and 3	0.150 m
lower limiting frequency f_l of the tube	200 Hz
upper limiting frequency f_u of the tube	3000 Hz

[a] *Distance can be changed by insertion of separate test sample holder*

The air tightness of the tube under microphone holders was ensured by gluing. The rubber seal arround the microphone body is also used to determine the correct insertion depth into the holder (Figure 3a). Microphones are flush mounted in their positions (Figure 3b).

a) 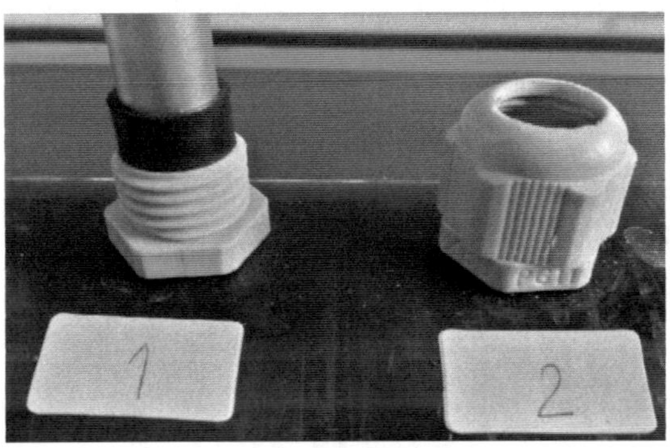

b) 

Figure 3. Design of mic holders (a) and details of the correct microphone installation (b).

Both parts of the impedance tube are joined together by screws. The second opposite tube has the same dimension parameters as the measurement tube. It contains the rigid piston and test samples inserted into this tube (Figure 4a). From the displacement of the piston, the sample thickness s_{thk} can be checked after insertion and/or depth of the air cavity in the back of the test sample (Figure 4b).

a) 

b) 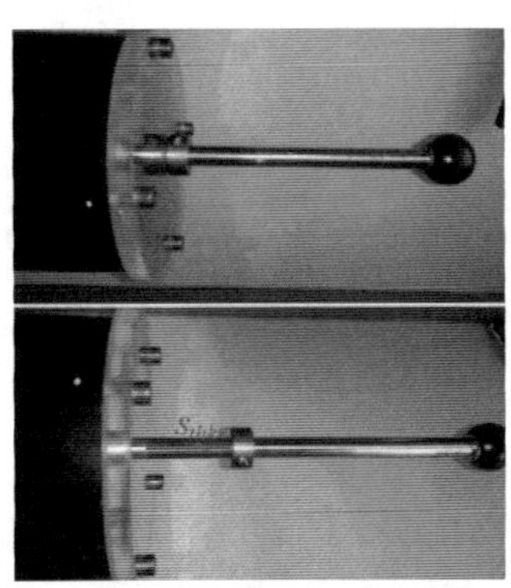

no sample inserted in the tube

inserted sample in the tube with thickness s_{thk}

Figure 4. Test sample installation.

The sound source is a loudspeaker mounted in a sealed box combined with a signal amplifier based on audio chip TDA2030 in the recommended circuit connection. The final design of the impedance tube prototype shows Figure 5. In the next work piezo-exciter as the sound source will be tested.

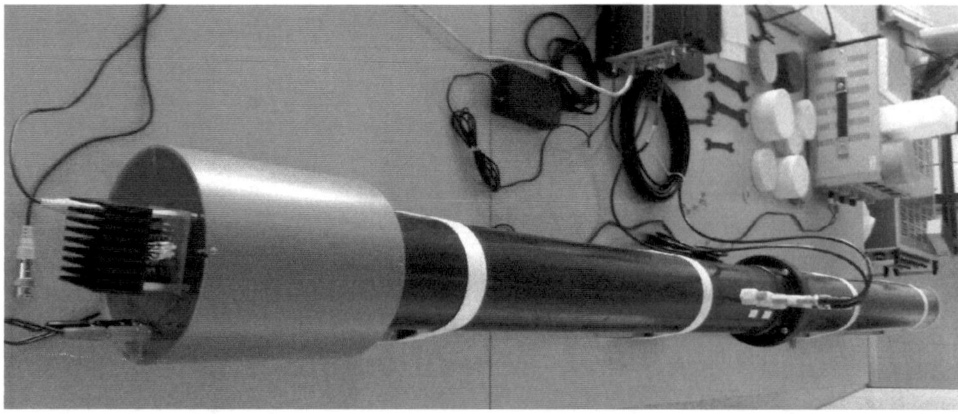

Figure 5. Prototype of the complete impedance tube.

5. Determination of natural frequencies of the impedance tube

The general way to identify the modal parameters of the structure is fitting a linear model by the Frequency Response Function (FRF), a function related to the excitation frequency spectrum $F(f)$ to the response frequency spectrum $X(f)$

$$X(f) = H(f)\,F(f) \tag{13}$$

Instrumentation used for the determination of natural frequencies of made impedance tube: hardware unit Brüel & Kjær FFT Analyzer Type 3560-B-120, four accelerometers Brüel & Kjær Type 4508-B, and modal hammer Brüel & Kjær Type 8206-001 with plastic tip. The accelerometers were installed at the tube using beeswax.

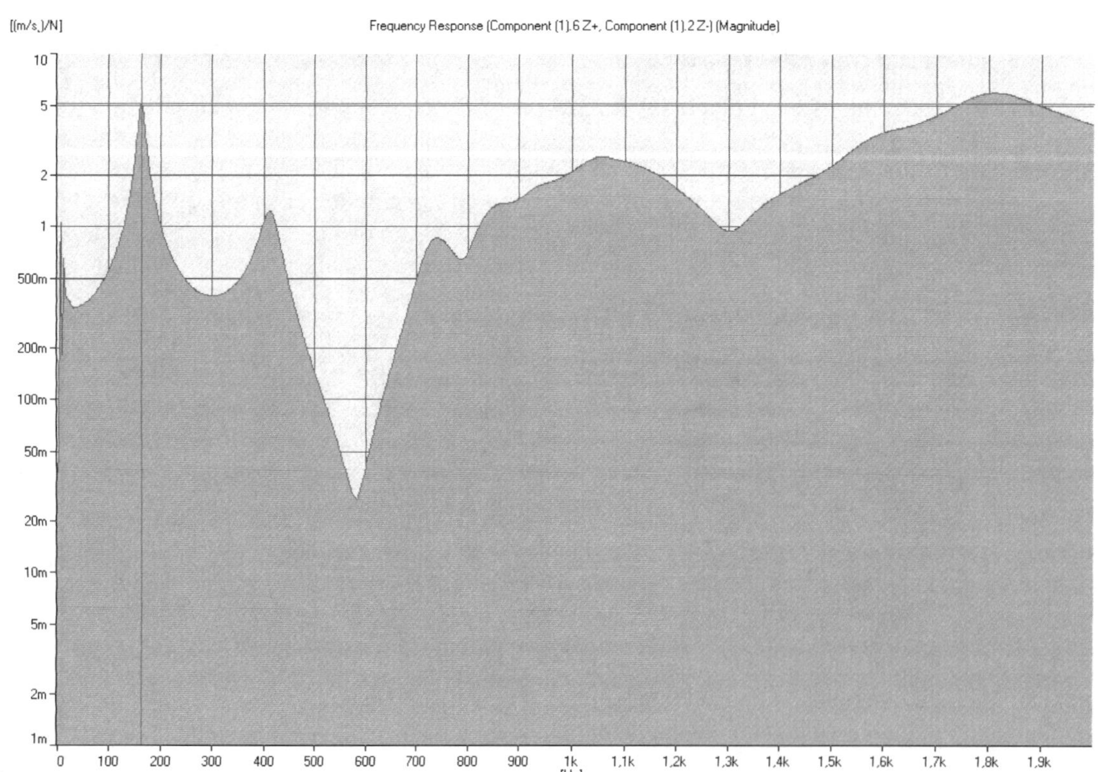

a)

IOP Conf. Series: Materials Science and Engineering 1050 (2021) 012003 doi:10.1088/1757-899X/1050/1/012003

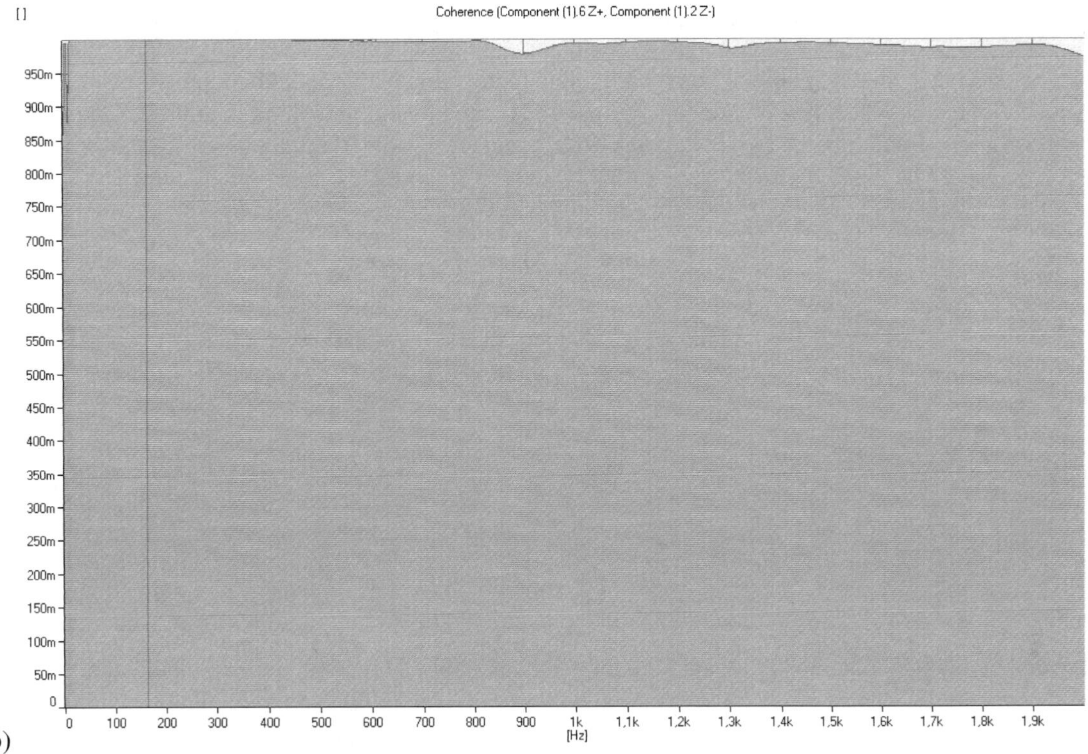

b)

Figure 6. Determination of modal parameters of the impedance tube (a) FRF and (b) the coherence function.

In Figure 6a, magnitude of frequency response function obtained by accelerometer placed near the microphone holder at position 2 is presented. Coherence function in Figure 6b indicates degree of linear relationship between the input and response signal at the working frequency range of the tube. From the experimental results of the modal analysis it followed that the 1st natural frequency was 164 Hz, under the range of interest for acoustic measurement, but the 2nd natural frequency 411 Hz with lower amplitude was near theoretical lowest limit frequency of the tube.

6. Measurement of the sample acoustic properties

The following equipment was used for the measurement based on the transfer function method: hardware unit Brüel & Kjær FFT Analyzer Type 3560-B-120, two microphones Brüel & Kjær Type 4189-A-021, built impedance tube with a signal amplifier, software Brüel & Kjær Pulse Labshop and as the source of white noise PC audio card was used. Acoustic material properties of foam were determined in the frequency range from 200 Hz to 1600 Hz. The sample was cut using a laser graving and cutting machine. The parameters of the test sample are summarized in Table 2.

Table 2. Parameters of the test specimen.

Parameter	Value
diameter d of the sample	0.088 m
thickness s_{thk} of the sample	0.044 m
material	foam
density ρ	28.25 kg.m^{-3}

IOP Conf. Series: Materials Science and Engineering 1050 (2021) 012003 doi:10.1088/1757-899X/1050/1/012003

All the measured data were processed by Pulse LabShop software. Figure 7 shows the results of acoustic material properties measurement during the test of impedance tube. Variation of the sound absorption coefficient (Figure 7a), the reflection factor (Figure 7b), and real part of the specific acoustic impedance ratio (Figure 7c) of the foam test samples are presented in following graphs. The sound absorption coefficient curve measured for foam of 44 mm thickness has similar tendency as results presented in [9] for the foam with thickness 40 mm. The constructed tube shows certain discontinuity of sound absorption coefficient and reflection factor around the frequency 850 Hz.

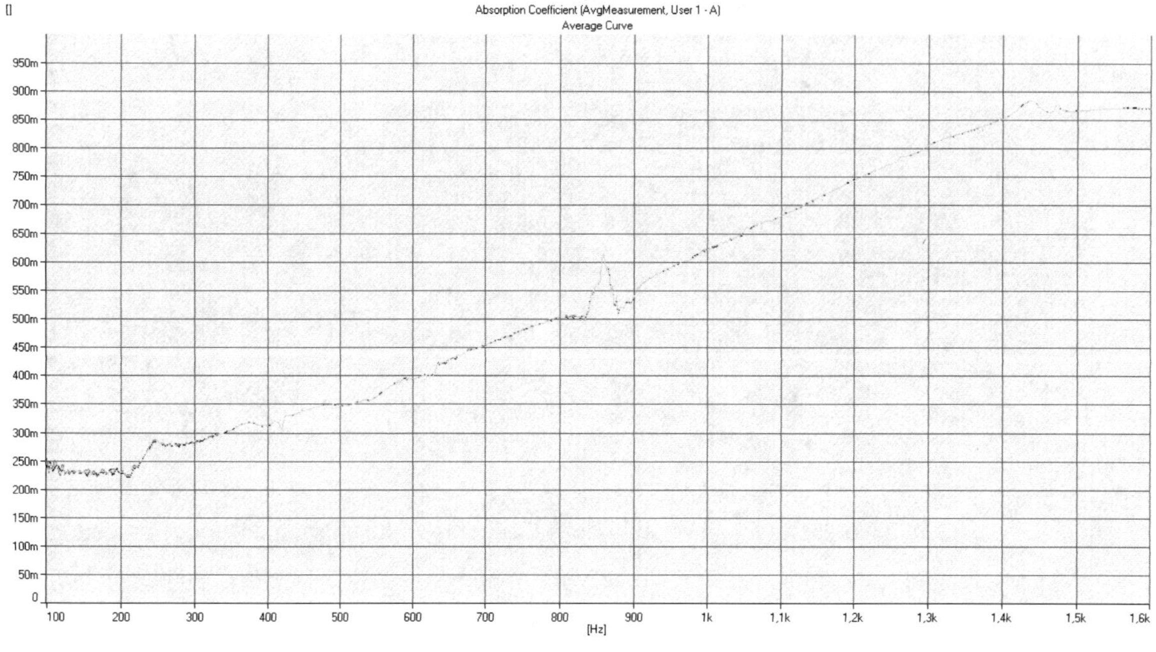

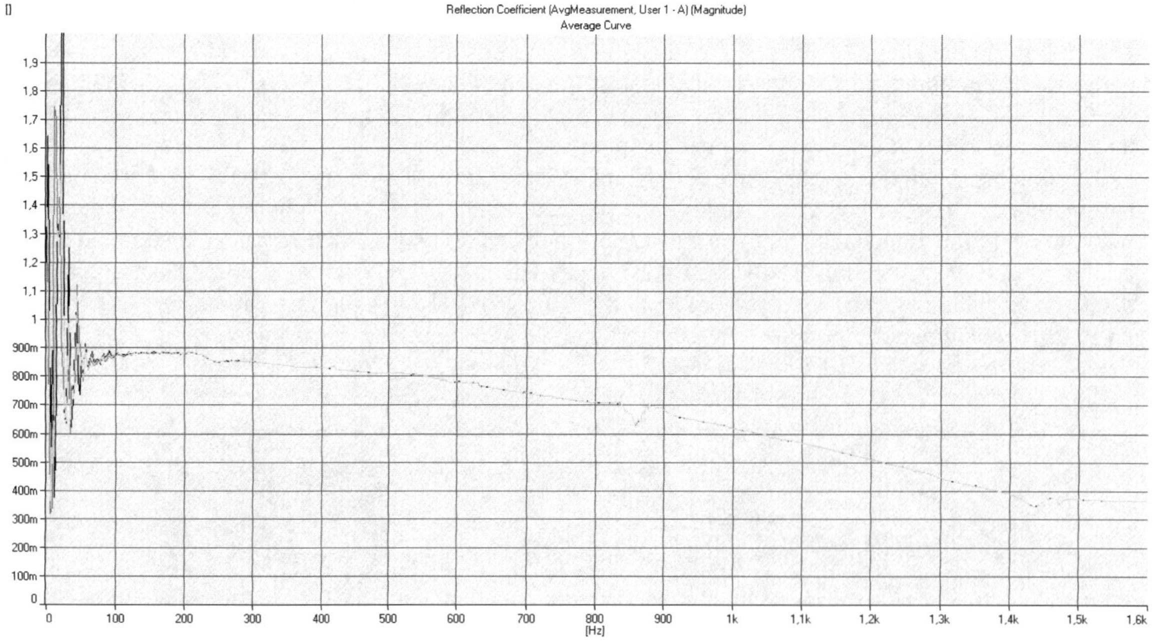

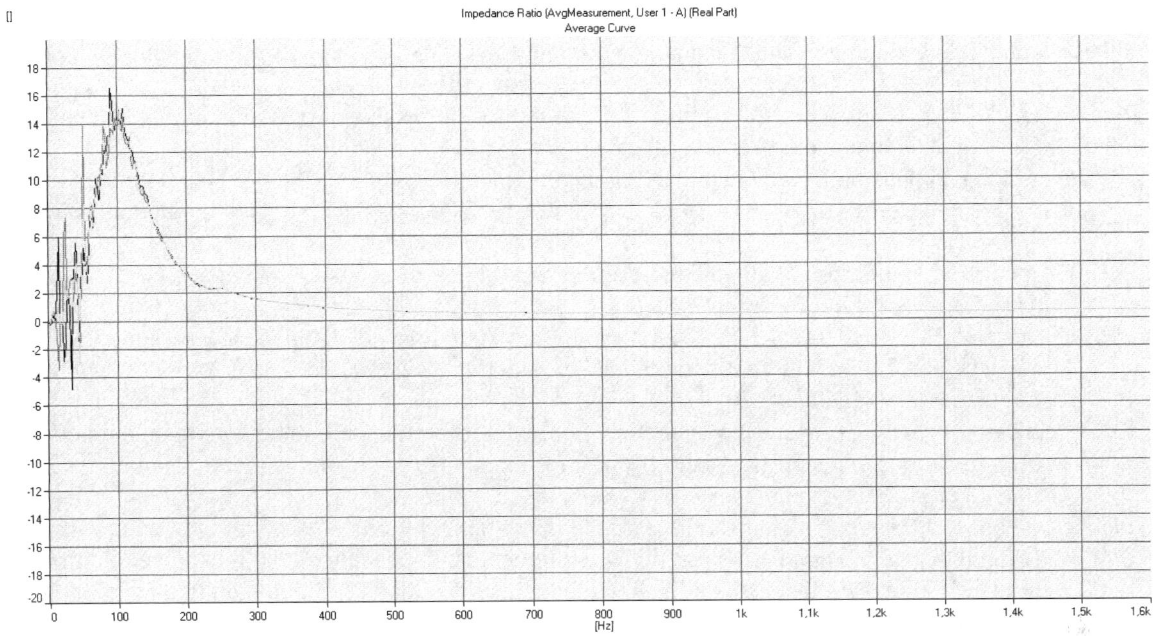

Figure 7. Frequency dependence of sound absorption coefficient, reflection factor and acoustic impedance ratio (real part) of tested foam material.

7. Summary

The design of the low-cost impedance tube is presented in this article. The prototype of the tube can be used for testing of acoustic properties of material in accordance with the mentioned ISO and ASTM Standards. The tested sample demonstrated approximately a linear rise of the sound absorption coefficient with the frequency in the range of interest. The reflection factor decreased its value in dependence on the frequency. In our future research, we plan to compare the acoustic properties of different materials measured by our impedance tube with the results obtained by a certified authority.
The measurements performed by new impedance tube so far show similar results to those available in the literature for equivalent materials. It can be concluded that the transfer function for measuring the normal incidence sound absorption coefficient is more effective as the method of standing wave ratio and application of discrete frequencies. It is expected that the certain drops in curves of acoustic coefficients can be improved using a more rigid tube, smaller microphones (1/4"), second tube with smaller diameter and different broadband excitation signals.

Acknowledgement

This paper has been supported by the KEGA - Cultural and Educational Grant Agency of the Slovak Republic under the grant No. 029STU-4/2018.

References

[1] ISO 10534-1 1998 Acoustics - Determination of sound absorption coefficient and impedance in impedance tubes - Part 1: Method using standing wave ratio.
[2] ISO 10534-2 1998 Acoustics - Determination of sound absorption coefficient and impedance in impedance tubes - Part 2: transfer-function method.
[3] ASTM E1050 1998 Standard test method for impedance and absorption of acoustical material using a tube, Two microphones and a digital frequency analysis system.
[4] Available online at https://www.bksv.com/-/media/literature/Product-Data/bp1039.ashx.
[5] Available online at http://www.vibro-acoustic.com (Application Note: AN9020B-02/2015 Inside SCS9020B – Impedance Tube).
[6] Available online at https://www.holmarc.com/impedance_tube_apparatus.php.

[7] Available online at https://www.mecanum.com/tubes-d-impedance-et-de-tr?lang=en.
[8] Available online at https://sagetechnologies.com/products/test-a-measurement/acoustical/impedance-tubes/63-spectronics-acupro-impedance-tube.
[9] Suhanek M, Jambrosic K and Domitrovic H 2008 Student project of building an impedance tube (Acoustics '08 Press Conferences, Paris) 4479-84.
[10] Wei-Hong T, Rosmaini A, Zunaidi N H, Daud R and Cheng E M 2015 Development of an Indigenous Impedance Tube *Appl. Mech. Mater.* **786** 149-55 doi:10.4028/www.scientific.net/AMM.786.149.
[11] Pasqual A M and Lara L T 2017 Time-domain simulation of acoustic impedance tubes, *J. Braz. Soc. Mech. Sci. Eng.* **39** (1) 67-79.
[12] Deshpande S P and Rao M D 2014 Development of low cost impedance tube to measure acoustic absorption and transmission loss of materials (*Conf. Proc. ASEE Annual Conference and Exposition, Indianapolis*).
[13] Deaconu M 2014 Comparative analysis of transfer function and standing wave methods in determination of acoustic absorption coefficient (*Proc. Conf. Sisom Acoustics 2014, Bucharest*) 22-23.
[14] Yunseon R 2003 The Acoustic Impedance Measurement System using two Microphones, Brüel & Kjær, Sound & Vibration Measurement A/S Denmark (Available online at http://www.sea-acustica.es/fileadmin/publicaciones/Sevilla02_phagen035.pdf).
[15] Bodén H and Abom M 1986 Influence of errors on the two-microphone method for measuring acoustic properties in ducts *J. Acoust. Soc. Am.* **79** (2) 541.
[16] Hassall J R and. Zaveri K 1988 *Acoustic Noise Measurements* (Brüel & Kjær, Denmark).
[17] Seybert A F and Ross D F 1977 Experimental determination of acoustic properties using two-microphone random-excitation technique *J. Acoust. Soc. Am.* **61** (5) 1362-70.

IOP Conf. Series: Materials Science and Engineering 1050 (2021) 012004 doi:10.1088/1757-899X/1050/1/012004

Moisture expansion of Ca-rich ceramic body

P Dvořáková[1], A Kloužková[1], M Kohoutková[2] and M Kolářová[1]

[1]Department of Glass and Ceramics, University of Chemistry and Technology, Prague, Technická 5, 166 28 Prague, Czech Republic;
[2] Central Laboratories, University of Chemistry and Technology, Prague, Technická 5, 166 28 Prague

Abstract. Calcareous ceramic bodies are typical for historical pottery, e.g. majolica and faience, as well as for contemporary production of ceramic tiles. The amount of the calcium component and the ratio of the other phases together with the firing temperature are responsible for the final properties of the ceramic body. The amount of the calcium component has a fundamental effect on the relative expansion of the ceramic body after firing and also during long-term usage or eventually during storage. The expansion behaviour is mainly associated with the planarity of the final product which is required in the case of large-format tiles. Expansion of the ceramic body is also related to the long-term stability of historical ceramics, especially glazed ceramic objects. In the case of low-fired porous ceramics moisture expansion of the ceramic body occurs due to a humid environment. It is a process of rehydration/rehydroxylation of unstable non-crystalline residues which remain from clay minerals in the ceramic body after firing. This process can be suppressed with a suitable mineralogical composition of calcareous ceramics. The main aim of this work is to study moisture expansion behaviour of calcareous ceramic bodies with a focus on both historical and contemporary ceramics.

1. Introduction

Natural raw materials used for traditional ceramic production (majolica, faience, utility earthenware, floor and roof tiles, etc.) contain a relatively high proportion of plastic clays. The basic component of clays are various types of clay minerals which are a source of aging when fired to a temperature of approx. 1100 °C, due to residues of clay minerals in dehydroxylated form [1]. The basis of aging process of rehydroxylation is influenced not only by the amount of reactive non-crystalline residues after the decomposition of clay minerals, but also by the second process, which is absolutely essential for this type of ceramic body, i.e. the decomposition of the calcium component. Calcium clays, crushed limestone, chalk, ground shells, marble dust, dolomite, wollastonite, calcium and soda-lime feldspars are the typical sources of calcium component in ceramic body [1, 2]. The temperature range for the decomposition of limestone, which is the most used raw material in our area, is closely related to the particle size and heating rate. Limestone decomposition can take place from 620 °C, while the main part of decomposition occurs in the range of 700–850 °C [1, 3-5]. During further temperature rise reactive calcium oxide is formed as a new crystalline phase. Furthermore, the calcium oxide reacts with metakaolinite to form gehlenite $2CaO.Al_2O_3.SiO_2$. Gehlenite as a thermodynamically unstable phase subsequently reacts with excess silica or other components of ceramic body (metakaolinite, spinel, etc.) to form a stable crystalline phase anorthite ($CaO.Al_2O_3.2SiO_2$) or other products such as secondary wollastonite ($CaSiO_3$) or diopside ($MgCaSi_2O_6$) [5-8]. If the new crystalline phases in the form of anorthite, gehlenite and mullite are formed during firing in the correct ratio (firing above 900 °C), it is possible to obtain a ceramic body with a lower value (almost half the value) of moisture expansion than that of a non-calcium ceramic body fired according to the same firing schedule [9-10]. The spontaneous process of rehydroxylation begins immediately after the firing of the ceramic body; in the beginning it is relatively fast, then it gradually slows down. However, it is always a long-term process, and the value of irreversible moisture expansion depends not only on the chemical and mineralogical composition of the raw materials, but also on the overall technological process of production.

The main aim of this work was to prepare a Ca-rich ceramic body with a low value of thermal expansion coefficient (CTE), which will be suitable for large-format tile production. Furthermore, the aging process of the prepared ceramic body was to be evaluated under the conditions of laboratory hydrothermal loading. A partial goal was focused on determining the source of deterioration of the glaze of historical faience. The aim of the work was to prepare a Ca-containing ceramics body with a low CTE suitable for large-format tiles and to evaluate its aging in the conditions of laboratory hydrothermal loading. The partial goal was to assess the source of damage to the glaze of historical faience.

Experimental part

Three model ceramic bodies (marked as 1, 2, 3), a contemporary ceramic body used for the production of tiles (marked as 0) and historical faience shard material (marked as F) were used to investigate the moisture expansion.

Experimental methods

Chemical compositions of powder samples were evaluated using a fully automatic sequential XRF spectrometer ARL 9400 XP (Thermo ARL, Switzerland). The X-ray fluorescence method was used for semi-quantitative and qualitative element analysis of up to 83 elements of the periodic table (B-C O-U) in the range from ppm to 100 %. The spectrometer is suitable for measurement of powders, solid materials, thin films and fluids and is equipped with a Rh tube, a 4kW generator, 4 collimators, 6 crystals (AX 20, TLAP, PET, Ge 111, LiF 200, LiF 220) and 2 detectors – proportional and scintillation. The data obtained were evaluated by the software Uniquant 4.

Mineralogical compositions were identified by X-ray diffractometer PANanalytical X´Pert PRO (PANanalytical, The Netherlands) equipped with a Cu lamp and 1D detector PIXCEL with variable divergence and detected aperture. Measured data were evaluated by the software X´Pert HighScore Plus 3.0 with a reference sample database.

Mineralogical composition of ceramic bodies was also investigated using attenuated total reflectance (ATR) method with multi-reflecting diamond crystal (Nicolet iS50 spectrometer). Data were collected by an Omnic software in the range between 1800 and 400 cm^{-1} with the spectral resolution 4 cm^{-1}.

The thermal expansion of the ceramic bodies was measured by LINSEIS L75 HS 1600C PT dilatometer. The heating rate was 5 °C min^{-1} up to 700 °C with 8 min dwell at the final temperature in helium and air flow. The data obtained were evaluated by the software Linseis Platinum Evaluation V1.0.90. Measurements were carried out on samples before and after hydrothermal loading. Fired compact samples of ceramic bodies were exposed to hydrothermal loading in a laboratory autoclave A/650/20 (Ceramic Instruments, Italy). The autoclave is built according to PED standards with a maximum pressure of 20.0 bar (2000 kPa), a manometer with a range of 0.25 bar (2.5 MPa) and an operating temperature of 230 °C.

Powder samples of ceramic bodies were measured also by simultaneous thermal analyses STA PT 1600/1750 °C HiRes (Linseis, Germany) in TG-DTA mode in a temperature range 25–1200 °C with a heating rate 10 °C min^{-1} in helium flow. The measured sample was placed in a platinum crucible. The released gasses, mainly CO_2 a H_2O, were analysed by ThermoStar^{TM} GSD320 (Pfeiffer Vacuum, Austria) mass spectrometer. Evaluation of measured data was performed using Linseis Evaluation software.

2. Results and Discussion

Model ceramic bodies (1, 2 and 3) were developed to study the dehydroxylation/rehydroxylation process of the Al_2O_3-SiO_2-CaO ceramic system. Ratios of the raw components were based on previous research of historical (F) and contemporary (0) Ca-rich ceramic bodies. Raw materials were combined in order to achieve a suitable mineralogical composition in the resulting model systems, which shows a delay in the area of small length changes during firing. Samples of model mixtures were fired at different final temperatures. The work was focused on the behaviour of model samples during temperature load and the formation of new phases in the ceramic body during firing. A combination of several analytical methods, mainly simultaneous thermal analysis (STA), X-Ray diffraction (XRD and infrared spectroscopy by attenuated total reflection (ATR), was used for the study.

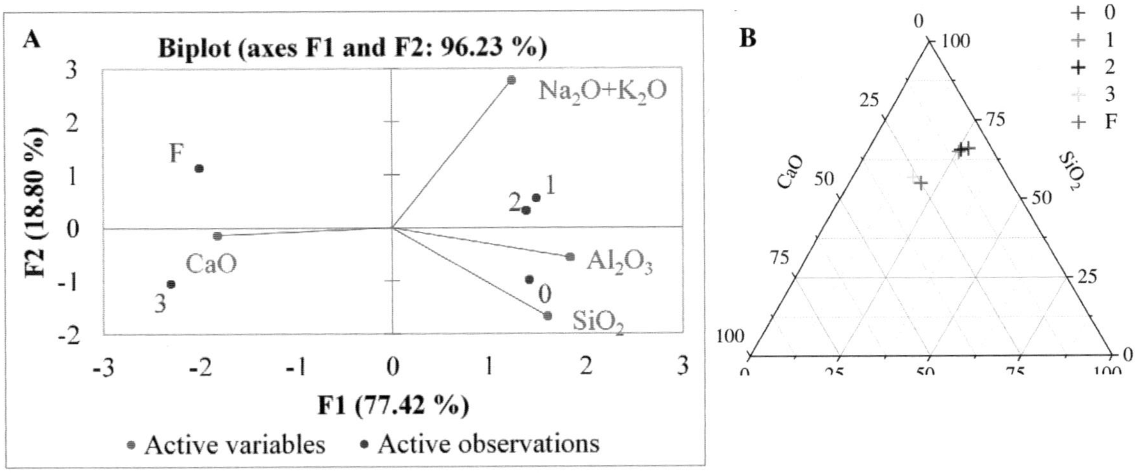

Figure 1 Chemical composition of studied ceramic bodies: (A) PCA biplot, (B) ternary diagram

The chemical composition of real and model ceramic bodies is shown in Figure 1 in the form of a PCA biplot, where the red lines are directives of the rising contents of main oxides. The chemical composition of the ceramic mixtures are presented also in the form of a ternary diagram (right hand side on the top) and in the bottom the Al_2O_3-SiO_2-CaO phase diagram is shown. Quartz and anorthite (calcium feldspar) were identified using XRD as the main mineralogical components. Diopside was identified in the historical faience sample (F) and wollastonite in the model sample (3) with a higher proportion of calcium component, which is presented in Figure 2.

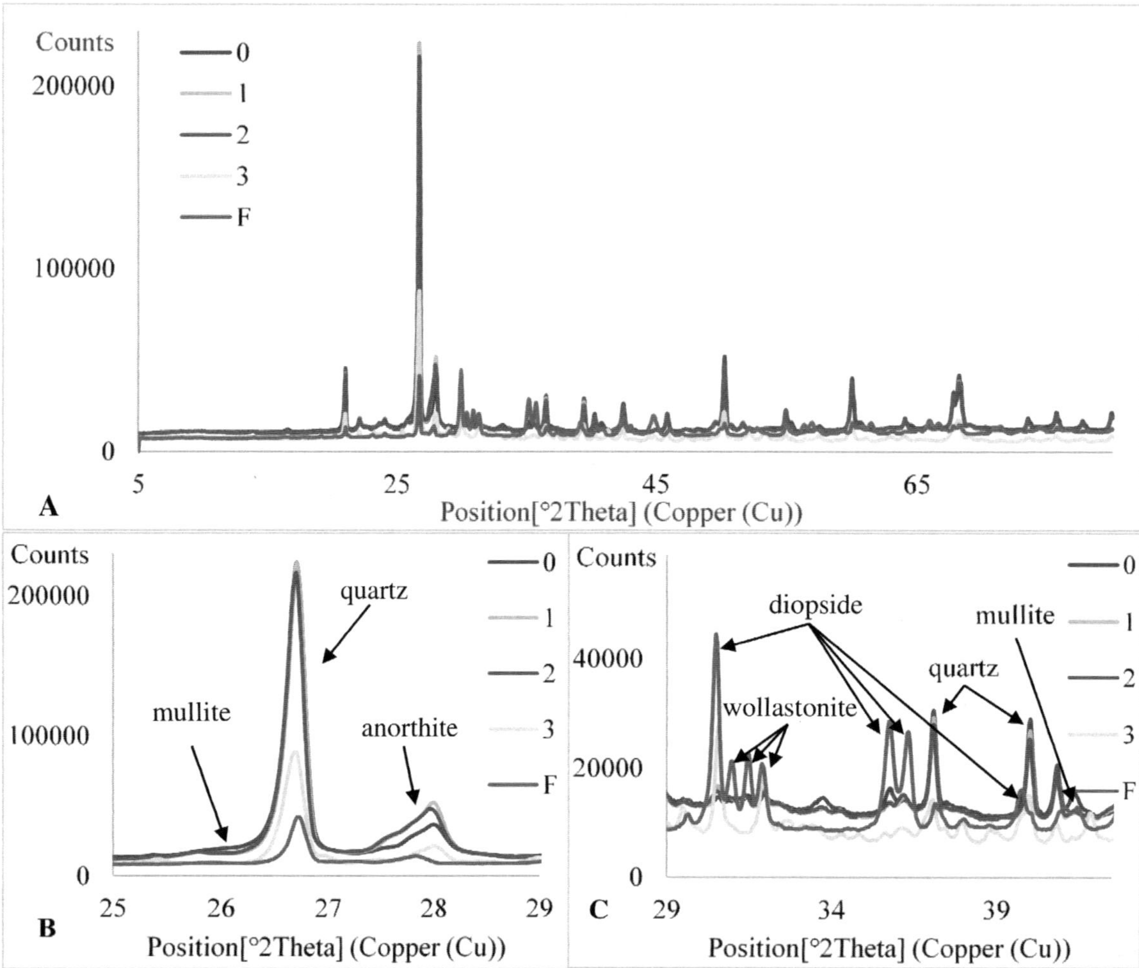

Figure 2 X-ray patterns of ceramic bodies (A) and two details of X-Ray patterns in the range of 25–29 °2θ (B) and 29–42 °2θ (C)

The ratio of anorthite was studied on fired model samples and it was found that the highest amount of anorthite was formed in the model sample (2) and the lowest in the sample (0), in which anorthite is supplemented by a higher proportion of mullite (see the detail of the X-ray pattern in Figure 2B). In the sample (0) also a trace amount of gehlenite was identified. In terms of mineralogical composition, model mixture (2) is the most resistant to deformation during the second part of firing (the area of small length change approx. 950–1160 °C). The resistance to deformation of model sample (2) was confirmed by dilatometric measurements in the temperature range 25–1200 °C (shown in Figure 4). Both anorthite and wollastonite were formed after firing in model sample (3), in which a higher content of CaO was identified by XRF (approx. 20 wt%). Diopside was identified as a main mineral in the historical sample (F). Furthermore, a small amount of gehlenite was determined in the mineralogical composition of historical sample.

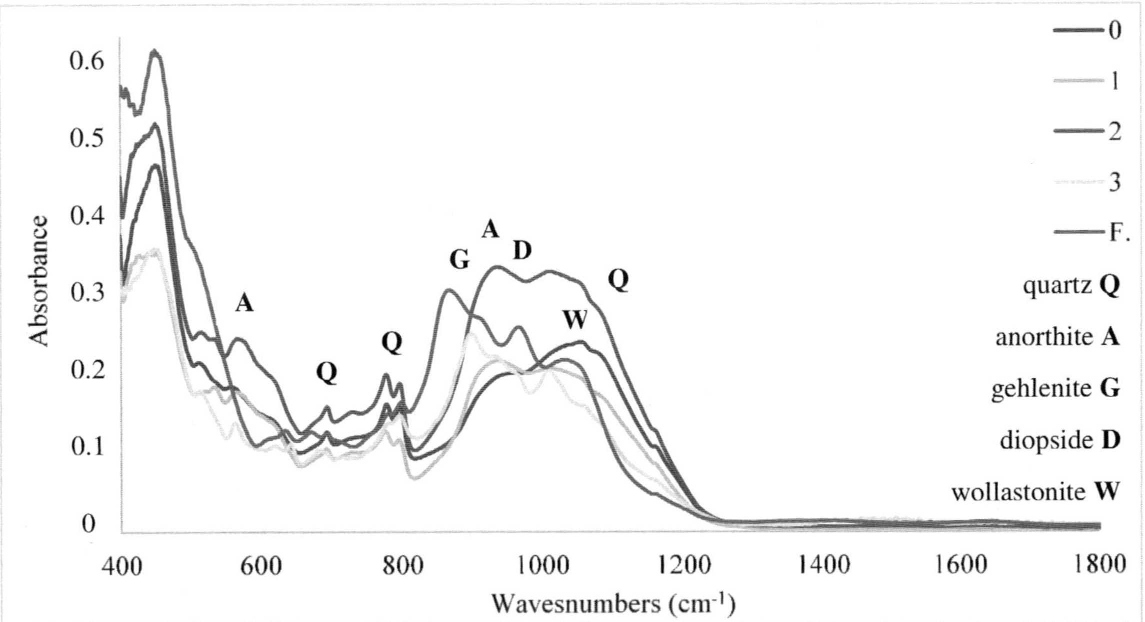

Figure 3 ATR spectra of samples of fired ceramic bodies

Figure 3 shows the measured ATR spectra of model and real ceramic samples, in which the main bands of identified minerals (1079, 796, 777, 694 cm^{-1} - quartz; 912 cm^{-1} - gehlenite; 571 and 942 cm^{-1} - anorthite; 955 cm^{-1} - diopside, 967 cm^{-1} - wollastonite) are indicated. The position of the Si-O vibration in the 1000–900 cm^{-1} region was characteristic for calcium minerals (gehlenite, anorthite, diopside, wollastonite). It was proved that IR spectroscopy could be used for identification of particular Ca-minerals. Decomposition of carbonates and CO_2 release from the raw model mixtures during firing process was studied by STA and MS. The DTA curves of all three model systems shows a similar course of carbonate decomposition in the temperature range 630–850 °C that corresponds to the CO_2 release identified by MS, because the particle size distribution of the Ca-rich raw material used was similar in all cases (Table 1).

Table 1 DTA parameters of model ceramic bodies

	T_{onset} (°C)	T_{offset} (°C)	T_{min} (°C)	Area (μVs mg^{-1})
0	631.7	738.8	715.5	15.74
1	654.1	758.4	740.9	25.06
2	659.5	757.7	742.1	23.68
3	654.7	855.7	781.0	56.87

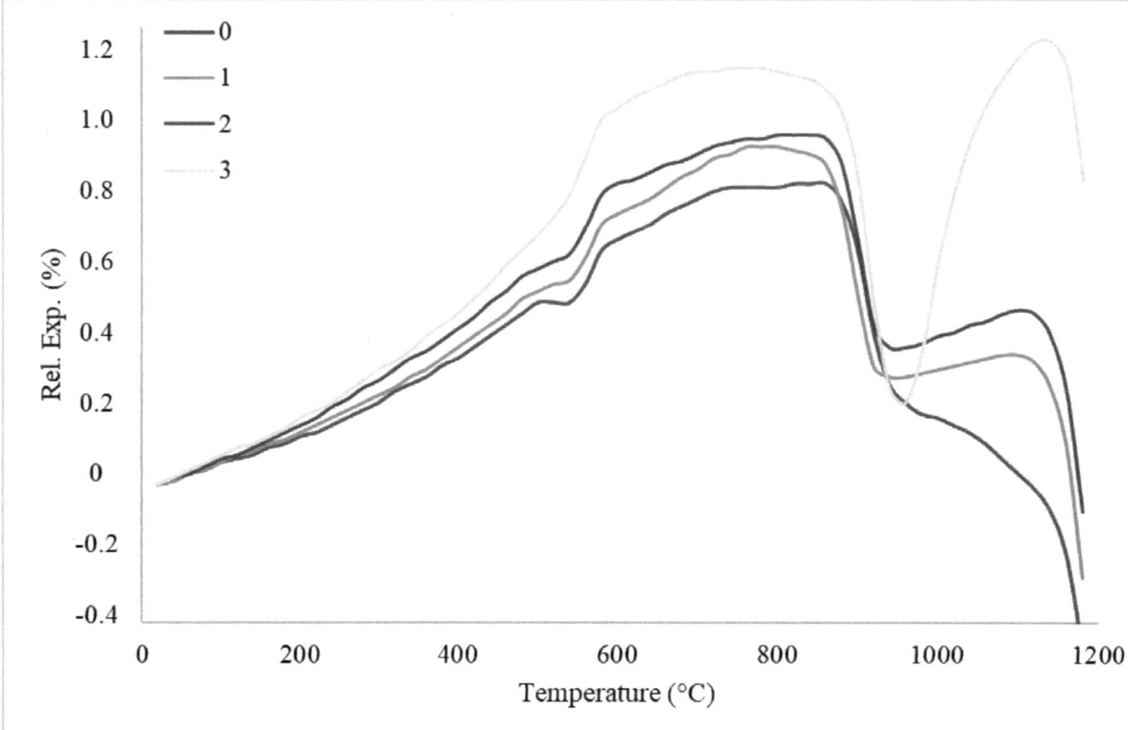

Figure 4 Thermal expansion curves of compact samples

A combination of several subsequent processes occurs during the firing of Ca-rich ceramic bodies. In the temperature range 700–850 °C different processes are involved (still ongoing dehydroxylation of clay minerals and onset of mica dehydroxylation, carbonate decomposition, and onset of gehlenite crystallization). The expansion of the ceramic body takes place in this temperature interval. Initial sintering and significant shrinkage occur above 850 °C. The area of small length changes (approx. 950–1160 °C) is characterised by crystallization of stable phases of anorthite and mullite. Shrinkage of the body is interrupted in the region of small length changes in the case of predominant anorthite phase (sample 1 and 2), which is shown in Figure 4. A significant increase in the volume of the model sample (3) is due to the crystallization of gehlenite and wollastonite. On the other hand, no noticeable interruption of shrinkage is detected in sample (0), in which the crystallization of mullite is dominant. The course of the firing process with regard to the formation of new phases was observed in sample (2). It was proved that with increasing firing temperature, there is a gradual increase in the anorthite content in the ceramic body. Model samples of the model mixture (2) were fired at different final temperatures (900, 950, 1000, 1050, 1100, 1150 and 1200 °C) using similar firing schedules, and part of the samples were subsequently exposed to hydrothermal aging. Dilatometric curves of samples before and after hydrothermal aging were measured and compared. Values of moisture expansion were obtained. The model samples fired at 900, 950 and 1000 °C contained residual metaclays, which rehydroxylated due to hydrothermal treatment (4.69 mm m^{-1}, 3.71 mm $m^{-1}$and 0.37 mm m^{-1}, respectively). This effect compensates the presence of anorthite. The amount of anorthite phase increases with increasing firing temperature. Samples fired at 1050 °C and higher contain a sufficient amount of anorthite which led to a significant inhibition of the aging process of the Ca-rich body. The mineralogical composition of the historical sample (F) was different, and it was proved that the faience body caused the crazing of the glaze.

3. Conclusion

A Ca-rich ceramic body exhibiting favourable behaviour in the regions of small length changes (approx. 950–1160 °C) and is suitable for the production of large-format tiles, was prepared. In the case of historical Ca-body, the effect of irreversible moisture expansion on the development of glaze defects was proved.

Acknowledgments
This work has been financially supported by the project of the Ministry of Culture Czech Republic DG18P02OVV028 "Technology of Treatment and Identification of Degradation Processes of Ceramic Finds from Hradčany Palaces – Methods of Restoration and Conservation of Porous and Dense Ceramics and Porcelain".

References

[1] Hanykýř V et Kutzendörfer J 2000 *Technologie keramiky* (Praha : Vega, s.r.o. ISBN 80-900860-6-3)
[2] Herainová M 2002 *Keramické suroviny a jejich úprava* (Praha : Silikátový svaz ISBN 80-903113-2-67)
[3] Escardino A, García-Ten J, Saburit A, Feliu C, Pilar Gómez Tena M 2013 Calcium carbonate decomposition in white–body tiles during firing in the presence of carbon dioxide *Ceramics International* **39**
[4] Siqueira F B et Holanda J N F 2018 Application of grifts waste as a renewable carbonate material in manufacturing tiles *Ceramics International* **4**
[5] Baccoura H, Medhioub M, Jamoussi F et Mhiri T 2009 Influence of firing temperature on the ceramic properties of Triassic clays from Tunisia *Journal of materials processing technology* **209**
[6] Traoré K T, Siméon K, Blanchart P 2003 Gehlenite and anorthite crystallisation from kaolinite and calcite mix *Ceramic International* **29**
[7] Taylor J R et Bull A C 1986 *Ceramics glaze technology* (Oxford : Institute of Ceramics ISBN 0080334660)
[8] Riccardi M P, Messiga B, Duminuco P 1999 An approach to the dynamics of clay firing *Applied Clay Science* **15**
[9] Sokolář R 2009 Vliv kalcitu na zatažení a nevratnou vlhkostní roztažnost cihlářského střepu. *Keramický zpravodaj* **25**
[10] Dvořáková P, Oujiří F, Kloužková A, Kavanová M, Kohoutková M 2019 Influence of composition on properties of a calcium ceramic body, *Proceedings of the XIIIth Internationl Conference* (Jahodná ISSN 978-80-533-3314-4)

The use of optimization in new design of tribometer specimen clamping system

Š Hajdu

Slovak University of Technology in Bratislava, Faculty of Materials Science and Technology in Trnava, Ulica Jána Bottu 2781/25, 917 24 Trnava, Slovak Republic

stefan.hajdu@stuba.sk

Abstract. The article deals with the innovation of the clamping part of the tribometer, where the tested specimens are clamped. The new clamping system was designed in the CAD system CATIA V5. The advantages of the new clamping system are compared with the original clamping system of specimens on the tribometer type A30. The new clamping design is evaluated from technical, technological and marginally also from economical point of view. The new clamping design ensures easy handling and exchange of specimens. Easy handling is ensured by the conical collet, which is released by the fixation nut, and subsequently the specimen can be pulled out and replaced with a new one without disassembly the device. The content of the article is also the strength analysis of the conical clamping collet and subsequent optimization. In the strength analysis, the conical collet is pushed into the conical cavity of the tribometer rotor by an outer fixation nut. The result is a numerical simulation of the new designed clamping system which will consist of the rotor, the conical collet, fixation nut and the tested specimen. The von Misses stresses, deformations and reaction forces acting on the collet will be determined by the numerical analysis. Subsequent optimization consisting of defining the target function, boundary conditions and type of optimization will be found. The minimum necessary fixation force which will be obtained from the fixation nut to ensure the transmission of torque to the clamped specimen without slip.

1. Introduction

When developing new devices, the main emphasis is on their reliability, durability, safety and ecology. These properties can be greatly improved by examining the friction and wear of mutually contacting surfaces that move relative to each other. Tribology deals with the study of these phenomena.

Tribological measurements are performed on specialized single-purpose devices, which task is to measure physical quantities arising from the interaction of two bodies moving relative to each other. In most cases, it is the temperature, the coefficient of friction and the pressing force. One of such devices is the Tribometer A30 manufactured by ZVL Dolný Kubín in 1981. This type of the device is currently being improved using CAD/CAE system CATIA V5. This software finds its application e. g. in the design of new machinery and equipment for areas such as engineering, automotive, aerospace, construction, agriculture industries and energetics [1].

Some packages of these computer programs also include structural analysis, thanks to which the designed devices can be easily verified in terms of strength, where even at the very beginning of the design it is possible to test selected structural parts. These parts are usually the most loaded member of machine parts. Structural analysis in commercial software packages is based on the finite element method. It is a modern numerical method for solving problems of continuum mechanics and is so far the most universal method known. The finite element method also solves physically nonlinear (plasticity) and geometrically nonlinear (large deformation, stability) problems. It usually solves such tasks that cannot be solved by classical methods [2-4].

Simulated annealing (SA) is a random-search technique which exploits an analogy between the way in which a metal cools and freezes into a minimum energy crystalline structure (the annealing process) and the search for a minimum in a more general system; it forms the basis of an optimisation technique for combinatorial and other problems. Simulated annealing was developed in 1983 to deal with highly nonlinear problems. Simulated annealing approaches the global maximisation problem similarly to using a bouncing ball that can bounce over mountains from valley to valley. It begins at a high "temperature" which enables the ball to make very high bounces, which enables it to bounce over any mountain to access any valley, given enough bounces. As the temperature declines the ball cannot bounce so high, and it can also settle to become trapped in relatively small ranges of valleys. A generating distribution generates possible valleys or states to be explored. An acceptance distribution is also defined, which depends on the difference between the function value of the present generated valley to be explored and the last saved lowest valley. The acceptance distribution decides probabilistically whether to stay in a new lower valley or to bounce out of it. All the generating and acceptance distributions depend on the temperature [5].

2. Problem description

2.1. Original clamping system of tribometer

A conical part made directly on the test specimen is pushed into the conical cavity of the tribometer rotor by means of a long screw (Figure 1). The torque from the rotor is transmitted to the test specimen by friction on conical surfaces. The conical surfaces also ensure the coaxiality of the rotor and the specimen. Such a solution requires a conical ending to be made on each tested sample. In order to ensure these properties, it is necessary that the design of these conical surfaces was accurate. For this reason, the production of precision conical surfaces is a demanding technological operation and therefore also expensive. The handling of the specimens in this setup is complicated and lengthy, as it is necessary to disassemble the entire load head and then pull the screw and the specimen out of the rotor to replace the specimen.

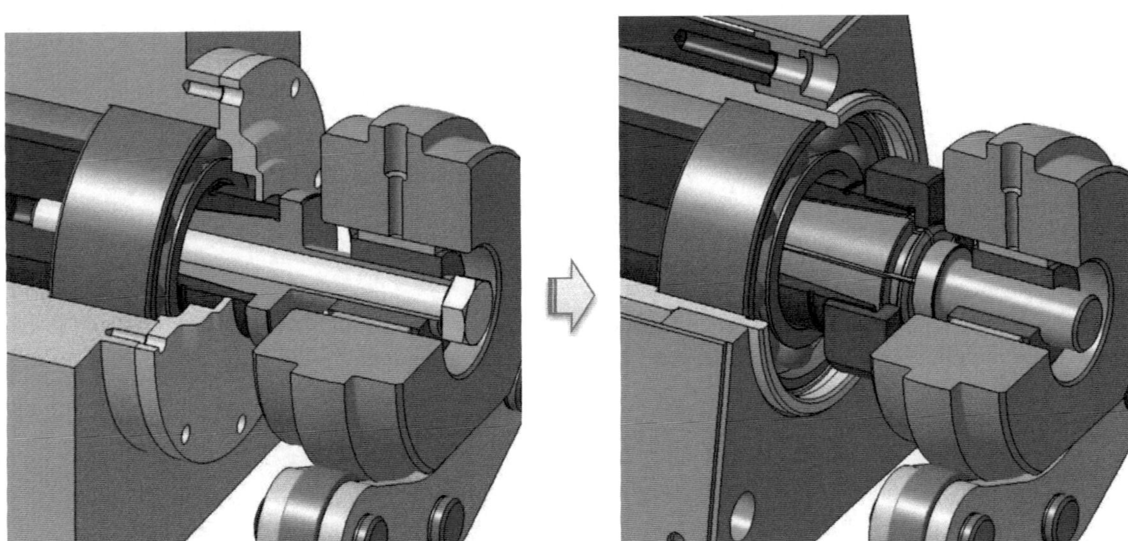

Figure 1. Original design with conical test specimen and screw fixation (left) and new design with conical-type of the spring collet (right).

2.2. New clamping system of tribometer

A conical clamping sleeve (collet) is pressed into the conical cavity of the tribotester rotor by means of a nut, into which the test specimen with a cylindrical end is clamped. By pushing the collet into the conical hole, the collet is compressed, thus forming a clamp connection between the cylindrical end of the sample and the inner cylindrical surface of the collet (Figure 1). The production of the samples has thus been significantly simplified, since it is not necessary to create a conical end on the samples, but only a cylindrical one. In this arrangement, the samples are exchanged by loosening the nut (green), thus released the collet (orange), and the sample can also be easily pulled out through the cavity of the loading head (pale brown).

3. Numerical simulation and optimization

The rotor, the collet and the specimen create a compression joint in cooperation with the nut. By pushing the conical collet into the conical cavity of the rotor, the conical collet tends to compress and thus reduce its inner diameter of the cylindrical hole (Figure 2). However, by inserting the cylindrical end of the specimen into the cylindrical hole of the collet a resistance is applied to this compression, which causes reaction forces to arise between participating parts of the assembly [4, 6].

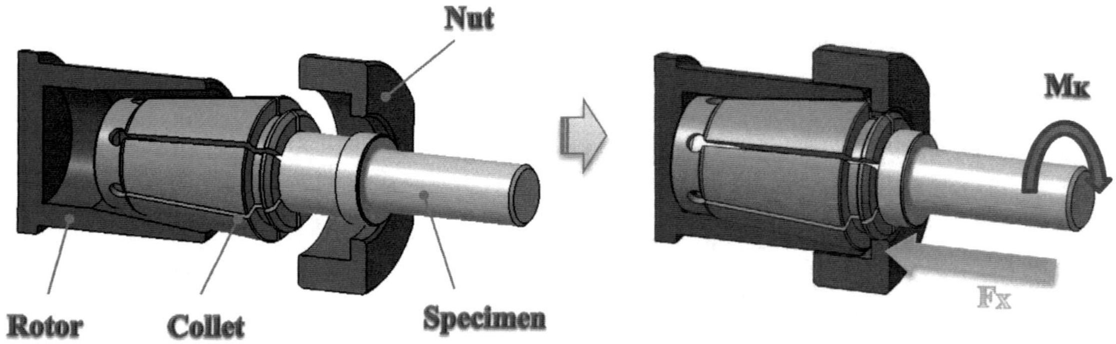

Figure 2. Considered state of the new clamping system for the numerical simulation.

3.1. CAD model

The conical collet chucks are standardized components and are manufactured in a variety of sizes and clamping ranges. It is therefore possible to clamp cylindrical specimens with different diameters by changing the collet without problems. In this case was chosen collet SN22 which the geometry and main dimensions are shown in Figure 3.

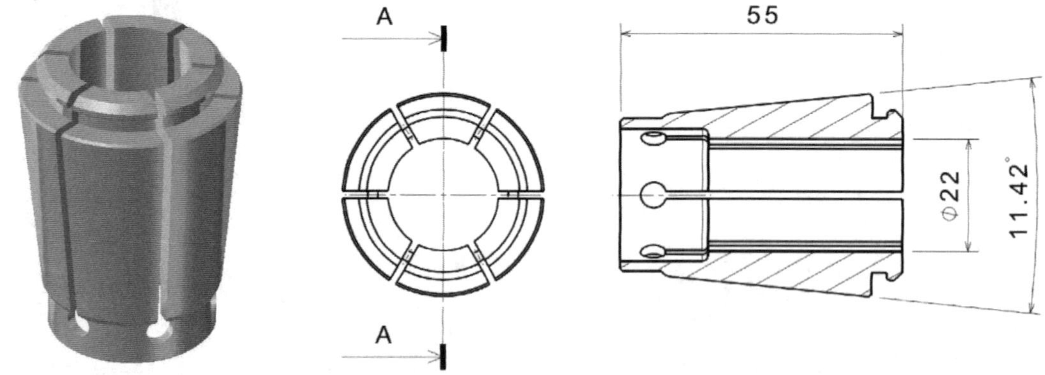

Figure 3. CAD model of the conical spring collet SN22 with main dimensions.

3.2. Finite element mesh

The maximal size of the element was set to 2 mm and absolute sag to 0.2 mm. The quality of mesh elements was checked and all generated elements were corrected in the whole volume of the part. In Figure 4, generated finite element mesh is shown, which contains of 28604 elements of the type TE10, 10-nodes parabolic tetrahedrons. The overall number of nodes for these mesh settings is 48372.

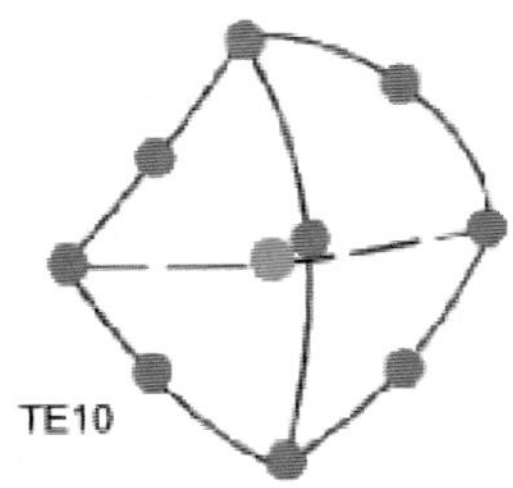

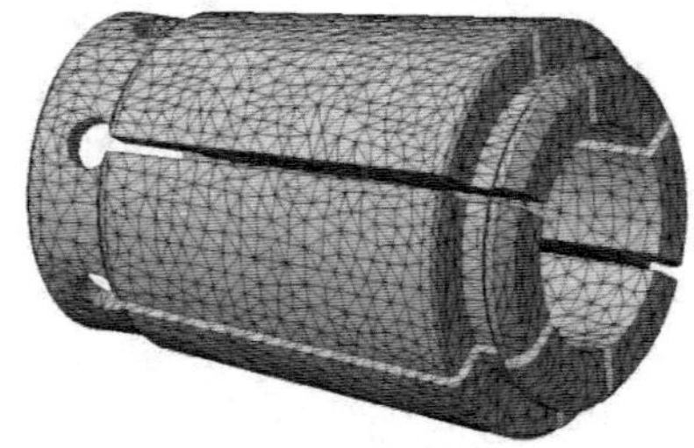

Figure 4. Parabolic tetrahedron element TE10 (left) and the finite element mesh generated using elements TE10 (right).

3.3. Material properties

In the simulation models, material properties of the collet were defined according to Table 1.

Table 1. Mechanical properties of the collet material.

Collet material	Young's modulus E [GPa]	Poisson's ratio ν [-]	Density ρ [kg.m^{-3}]
Steel	200	0.266	7860

3.4. Boundary conditions and loads

In the simulation, all interactions between each parts (Figure 2) were idealized by adequate boundary conditions [7-9] which can be seen in Figure 5. The outer and internal cylindrical surfaces have defined boundary conditions as surface sliders with no friction considered behaviour between the surfaces in contact. Surface sliders are surface constraint joins, which allow points of a surface to slide along a coinciding rigid surface. At each point of the deformable surface, the program automatically generates a constraint which fixes the translation degree of freedom in the direction normal to the surface at that point. The loading was defined as axial distributed force on surface where the contact is between the fixation nut and the collet.

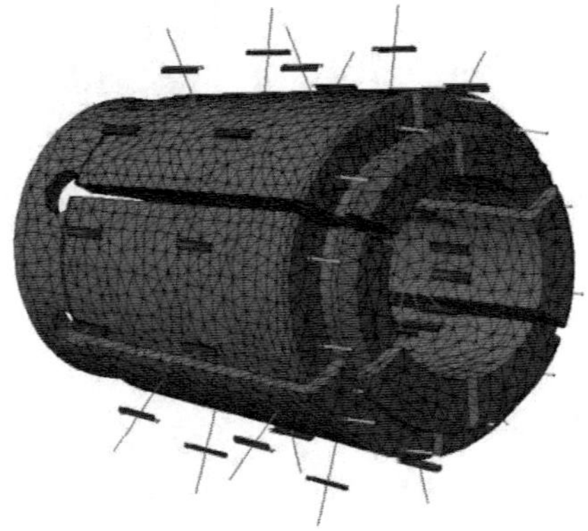

Figure 5. Boundary conditions and loads applied on the collet.

IOP Conf. Series: Materials Science and Engineering 1050 (2021) 012005 doi:10.1088/1757-899X/1050/1/012005

3.5. Type of the analysis
In this kind of numerical simulation, static type of the analysis has been chosen because the fixation nut creates constant load on the collet, and this fixation state is permanent during whole time of the tribological testing process.

3.6. Optimization settings
For the testing of specimens on the tribometer, it is necessary ensure the torque transmission to be reliably transmitted to the tested specimen. The construction of the tribometer is designed for the maximum torque value of M_K = 10 000 N.mm (Figure 6) which can be reached during testing. The safety factor $k = 2$ is also taken into account. It is clear from the construction of the collet that if we want to transfer such torque at the value of the friction coefficient for steel-steel contact $f = 0.3$, it is necessary that the adequate normal force F_N acting on the tested specimen will be created through a conical collet on its central cylindrical hole surface [5].

$$F_N = \frac{F_T}{f} k = \frac{M_K / r}{f} k = \frac{1000/11}{0.3} 2 \cong 6060 \text{N} \tag{1}$$

Since we already know the value of the required normal force F_N, it is possible to define it as the target value of the optimization function. The algorithm type of this optimization was set on simulated annealing. The axial force F_x which is found by this optimization has been defined as optimized parameter. Based on the change of this value, the value of the reaction force is calculated, which should be as close as possible to the value F_N = 6 060 N, which was calculated according to Equation 1.

3.7. Run optimization
Once we have all the necessary settings set, we can start the optimization process and collect the results in this stage of the numerical simulation (Figure 7).

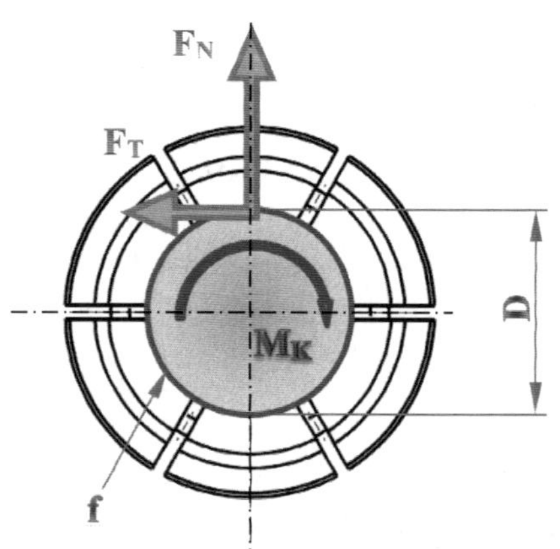

F_T - frictional (tangencial) force
F_N - normal force
f - coefficient of friction
M_K - torsional moment
r - internal collet radius

Figure 6. Schematic front view on forces and moments acting on the collet.

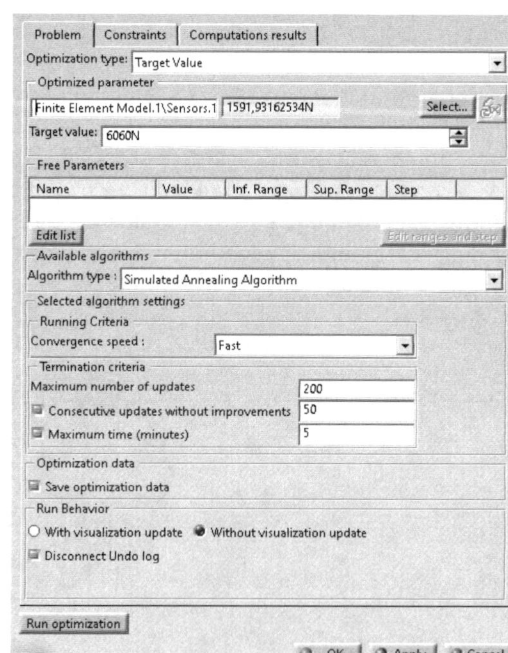

Figure 7. Detailed optimization settings.

IOP Conf. Series: Materials Science and Engineering 1050 (2021) 012005 doi:10.1088/1757-899X/1050/1/012005

4. Obtained results and discussion

The use of material is a good indicator for evaluating the benefits of a new method of clamping specimens. For the original method of the clamping system, the specimen is made of semi-finished product ø 45×110 mm. The weight of the semi-finished product is 1 375 g and the weight of the real part is 289 g. The use of material in this case is therefore only 21 % and the remaining 79 % is a waste (Figure 8).

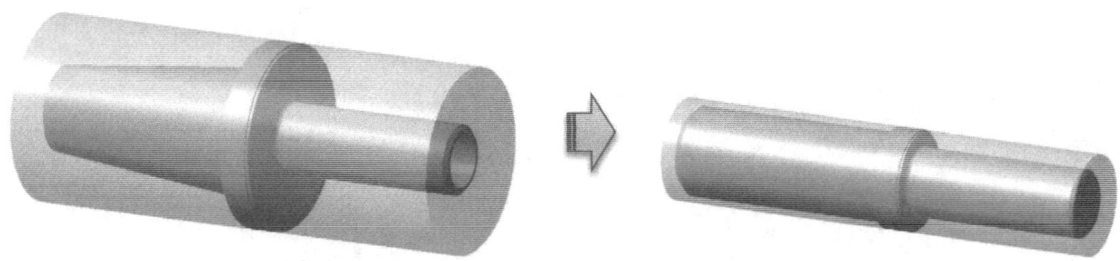

Figure 8. Original design of test specimen (left) and new design of test specimen (right).

In the case of a new design of clamping specimens, the specimen is made of the smaller semi-finished product ø 25×110 mm. The weight of the semi-finished product is thus only 424 g and the weight of the real part is 267 g. The use of material in this case is up to 63 % and only 37 % is a waste. Another advantage of the test sample is cheaper production, because grinding cylindrical surfaces is cheaper than grinding conical surfaces.

By optimization of the strength analysis parameters obtained the value of the force required to press the conical collet into the conical rotor hole by means of a nut in order to obtain the value of the normal force required to produce a frictional force capable of reliably transmitting torque from the rotor to the test specimen. Its value is approximately 3 800 N in the given conditions as can be seen from Figure 9. The minus sign before this value resulted from the geometric orientation of the model, where the direction of pushing was directed to negative values. Thus, the minus sign only determines the direction of action in a given coordinate system.

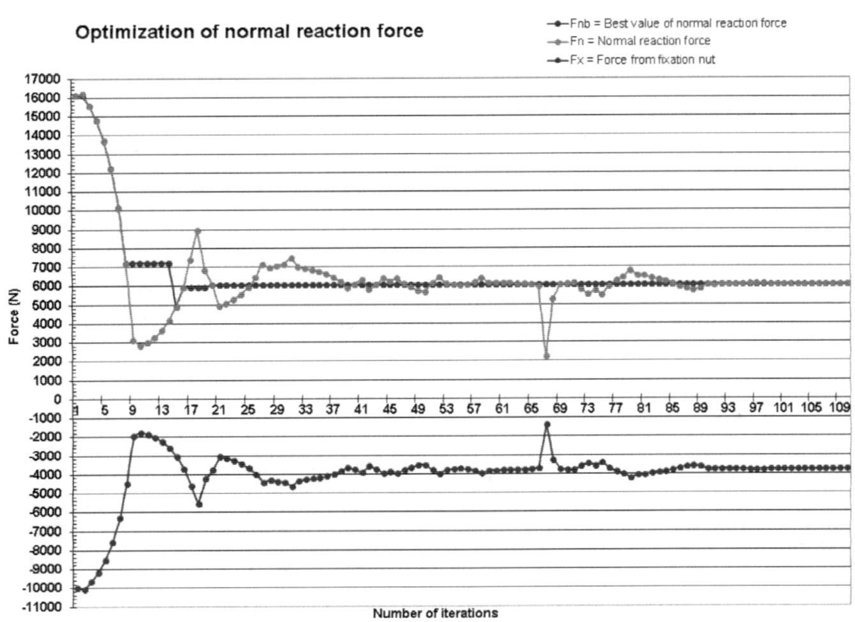

Figure 9. Results of optimization for 110 iterations.

At the peak value in 67th iteration is only checking value for finding global minimum value not only local minimum value of the finding parameter. In this case has been returned process of the calculation near the old results what signalize that this solution region is the best of all for finding the optimal solution of this problem. Under the action of such a large clamping force (approx. 3 800 N), the values of mechanical stresses (Figure 10) and displacements (Figure 11) for collet were calculated.

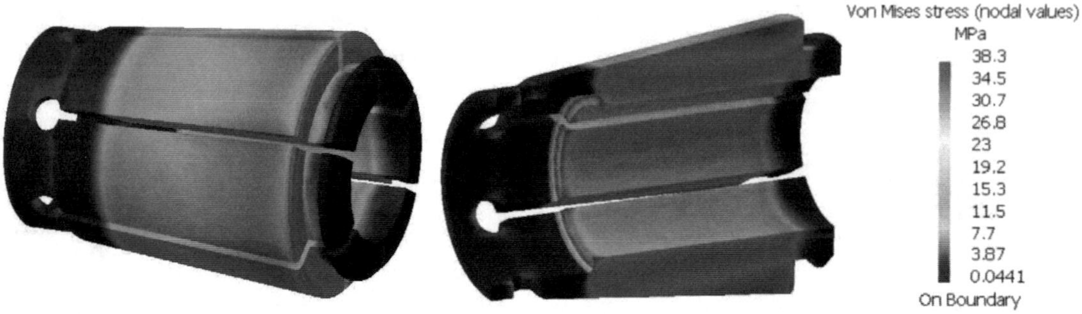

Figure 10. Calculated von Misses stresses in the conical spring collet.

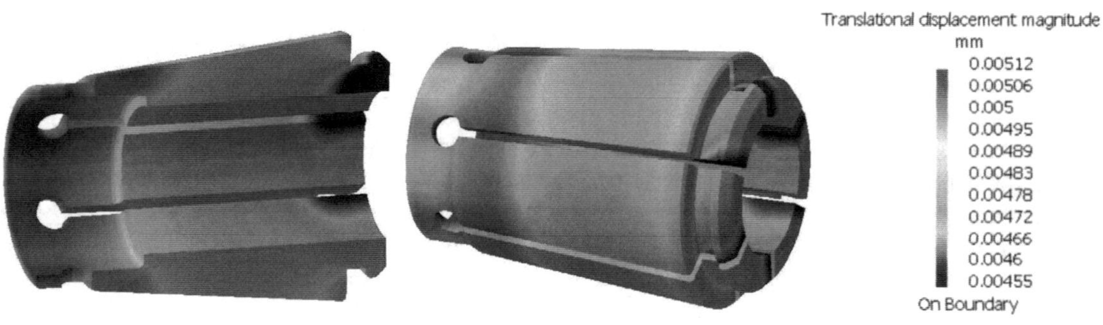

Figure 11. Calculated displacements on the conical spring collet.

5. Conclusions

The innovation of the tribotester clamping system has many advantages. The exchange of samples is simple and fast, because the loosening of the nut becomes an automatic loosening of the clamping system due to the pre-stressing of the conical collet. The collet is a standard part that can be replaced with a new one if it is damaged or lost. The semi-finished product from which equivalent types of new cylindrical samples can be produced represents only 30.9 % of the volume of the original conical shaped semi-finished product. Also, the use of material to make new cylindrical specimens is 63 % compared to the original conical design, which uses only about 21 %. Since the waste of material is minimized in a significant scale, it can be determined that not only from an economic point of view, but also from an ecological point of view, such a solution of clamping the specimen is more advantageous.

The value necessary to push the collet into the conical rotor hole must be 3 800 N to ensure a smooth and reliable transmission of torque to the shaft of the test specimen. Reliable torque transmission will only be ensured if the value of the coefficient of friction between the clamping sleeve pair is greater than 0.3. Also, the transmitted torque must not acquire a value 20 000 N.mm and the diameter of the shaft must not be less than 22 mm. When pressed with a force of 3 800 N, stresses are generated on the collet, which reach a maximum value of approximately 40 MPa. From the values of displacements, it is possible to derive a compressive force of 3 800 N necessary to push the collet into the conical rotor hole only 0.005 mm.

Acknowledgment
This contribution was written with financial support of VEGA Agency within the project 1/0232/18 "Using the methods of multi-objective optimization in production processes control".

References
[1] Spišák E 2000 *Mathematical modelling and simulation of technological processes* (Košice: TYPO Press Košice) ISBN 80-7099-530-0.
[2] Beroun S 2010 *Introduction to mechanical engineering* Available online at http://www.kvm.tul.cz/getFile/id:1886.
[3] Murín J 2010 *FEM in mechatronics* Available online at http://aladin.elf.stuba.sk/Katedry/KMECH/slovakversion/.
[4] Španiel M and Herák Z 2011 *Introduction to finite element method* (Praha ČVUT) ISBN 978-80-01-04665-4.
[5] Busetti F 2001 Simulated annealing overview Available online at https://www.researchgate.net/publication/238690391.
[6] Trebuňa F and Šimčák F 2005 *Elasticity, strength and plasticity in engineering* (Emilena Košice) ISBN 80-8073-276-0 (in Slovak).
[7] Herák D, Chotěborský, R Sedláček A and Janča E 2006 Exploitation of Hertz's contact pressures in friction drives *Res. Agr. Eng.* **52** (3) 107-114.
[8] Tököly P, Gajdoš M and Bošanský M 2009 *Effect of tooth shape to size of contact stress noninvolute gearing* In Vestnik Nacionaľnogo techničeskogo universiteta XPI 19/2009 Nacionaľnij techničnij universitet 10-20.
[9] Ďuriš R and Labašová E 2012 Experimental determination of the coefficient of friction in rotational sliding joint *Appl. Mech. Mater.* **309** 50-54 https://www.scientific.net/AMM.309.50.

Study of bovine serum albumin interaction with zinc oxide surfaces by force field simulations

H Hematian and B Rezek

Faculty of Electrical Engineering, Czech Technical University in Prague, Technická 2, 16227 Prague, Czech Republic

hemathad@fel.cvut.cz

Abstract. We studied the interactions of bovine serum albumin (BSA) molecule with ZnO surfaces by force field simulations. The different orientations of BSA over the two main nonpolar wurtzite ZnO surfaces $(10\bar{1}0)$ and $(11\bar{2}0)$ have been investigated. These surfaces contain both Zn and O atoms. We compare the results also with polar Zn-face (0001) surface. The results demonstrate tendency of BSA molecule for being adsorbed on the ZnO surface in specific orientations. This happens either by bonding between a hydrogen atom of BSA molecule to an oxygen atom of the ZnO surface or by physical attraction inducing ZnO atoms displacement in some cases.

1. Introduction

Serum albumin (SA) is the most plentiful protein in blood plasma and has been investigated intensively for its numerous vital functions such as maintaining the oncotic pressure, transporting drugs as well as contribution in functionality of the immune system. The interactions of SA with a wide range of nanomaterials have been the subject of many investigations [1,2]. The adsorption of SA on diamond as one of the most popular materials in biomedicine and electronics has been already investigated [3–6]. Besides diamond, zinc oxide (ZnO) as a wide and direct band-gap semiconductor with well-known optical properties, low level of toxicity and great biocompatibility possesses luminescence property that can be extremely useful in various biomedical and pharmaceutical applications [7]. The interactions of ZnO with a range of biomolecules and proteins have been reported by multiple experimental studies [8,9]. In [8], an *in vitro* toxicological investigation supported by theoretical modelling has shown multilayer adsorption of lactate dehydrogenase (LDH) on ZnO nanomaterials. Strong interaction of adenosine triphosphate (ATP) with ZnO nanostructures through chelate formations has also been demonstrated by Raman spectroscopy, XRD, and electron microscopy in [9]. In addition, using UV–vis absorption, fluorescence, synchronous fluorescence, and Raman spectroscopy to study of ZnO nanoparticles covalently and non-covalently coated by bovine serum albumin (BSA) has shown that the binding between ZnO nanoparticles and BSA decreases the cytotoxicity of ZnO nanoparticles [10]. Despite availability of large amount of experimental data conducive to perceiving the essence of ZnO-proteins interactions, fine aspects behind the interactions still need to be studied theoretically to understand the details of the phenomena on atomic scale.

However, the main challenge in computational study of the systems constituting nanomaterials and complex molecules such as albumin is the huge number of the atoms in the system which can make the calculations extremely costly if a proper approach is not adopted. One possible approach to cope with large systems is selecting more computationally efficient methods, which are still sufficiently accurate, over more costly ones with higher levels of accuracy (such as DFT and *ab initio* methods). Another complementary way can be considering a reasonable simplification in the system under study, such as fixing or excluding some non-interacting atoms in computations.

In this study, we aimed to elucidate the interaction of a BSA molecule with ZnO surfaces on atomic scale. Force field simulations have been carried out under some simplifying assumptions to overcome the computational limitations imposed by the huge size of the system. The interactions between BSA and ZnO for nonpolar and polar surfaces of ZnO have been compared.

2. Method

The adsorption of BSA on ZnO nonpolar surfaces which contain both Zn and O atoms has been investigated by implementing force field method in QuantumATK software which enables the possibility of dealing with the huge number of atoms in our system. The results have also been compared to polar Zn-face (0001) surface. HP system with a Core i5 8th generation processor and 8 GB RAM was used to perform the computations.

The system under the study consists of a BSA molecule and ZnO slab that are available from the RCSB Protein Data Bank website and QuantumATK software database respectively. Since hydrogen atoms are generally not observed in X-ray crystal structures, BSA structure taken from the Protein Data Bank lacks hydrogen atoms. Mol Probity has been employed to add hydrogen atoms to the molecule. Some approximations were also considered to provide the feasibility of using predefined force field potential sets as there are no sets of force fields defined for all types of atoms in BSA+ZnO system. Sulphur atoms have been excluded from the structure of BSA molecule to enable ReaxFF potential for force field simulations. This seems a reasonable simplification as sulphur atoms in BSA are not situated at or near the locations where the interaction of the molecule with ZnO occurs. To deduct the volume and duration of the calculations, different orientations of BSA molecule have been manually adjusted over ZnO surfaces in the distance range of 3-20 Å. The optimization trajectory and energy of the system then have been observed to discuss the interactions of the BSA molecule with ZnO planes.

3. Results and discussion

The simulations have been carried out for different orientations of BSA over nonpolar and polar ZnO planes. Figure 1 depicts five non-equivalent configurations used for forcefield simulations which are named based on the orientation of BSA molecule versus ZnO surface: Head, Side, Flat, Legs, and Tip. The distance between the nearest atom of BSA molecule and the surface of ZnO slab is set manually at around 3 Å for all configurations. The simulations have also been performed for the distance of 20 Å where there was no interaction between BSA molecule and ZnO surface detected regardless of the orientation of BSA.

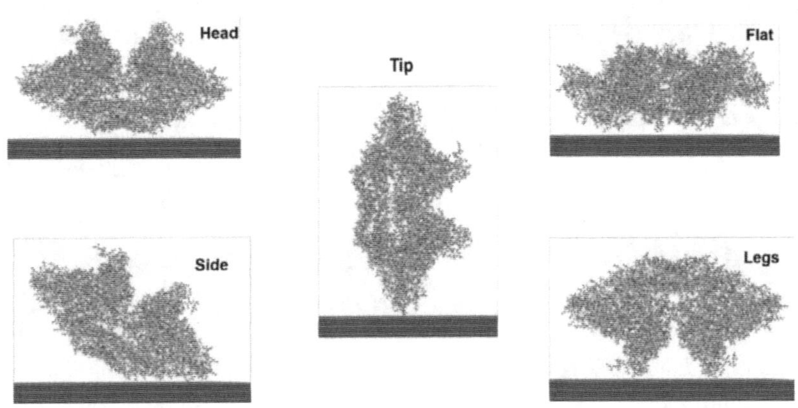

Figure 1. Different orientations of BSA molecule versus ZnO.

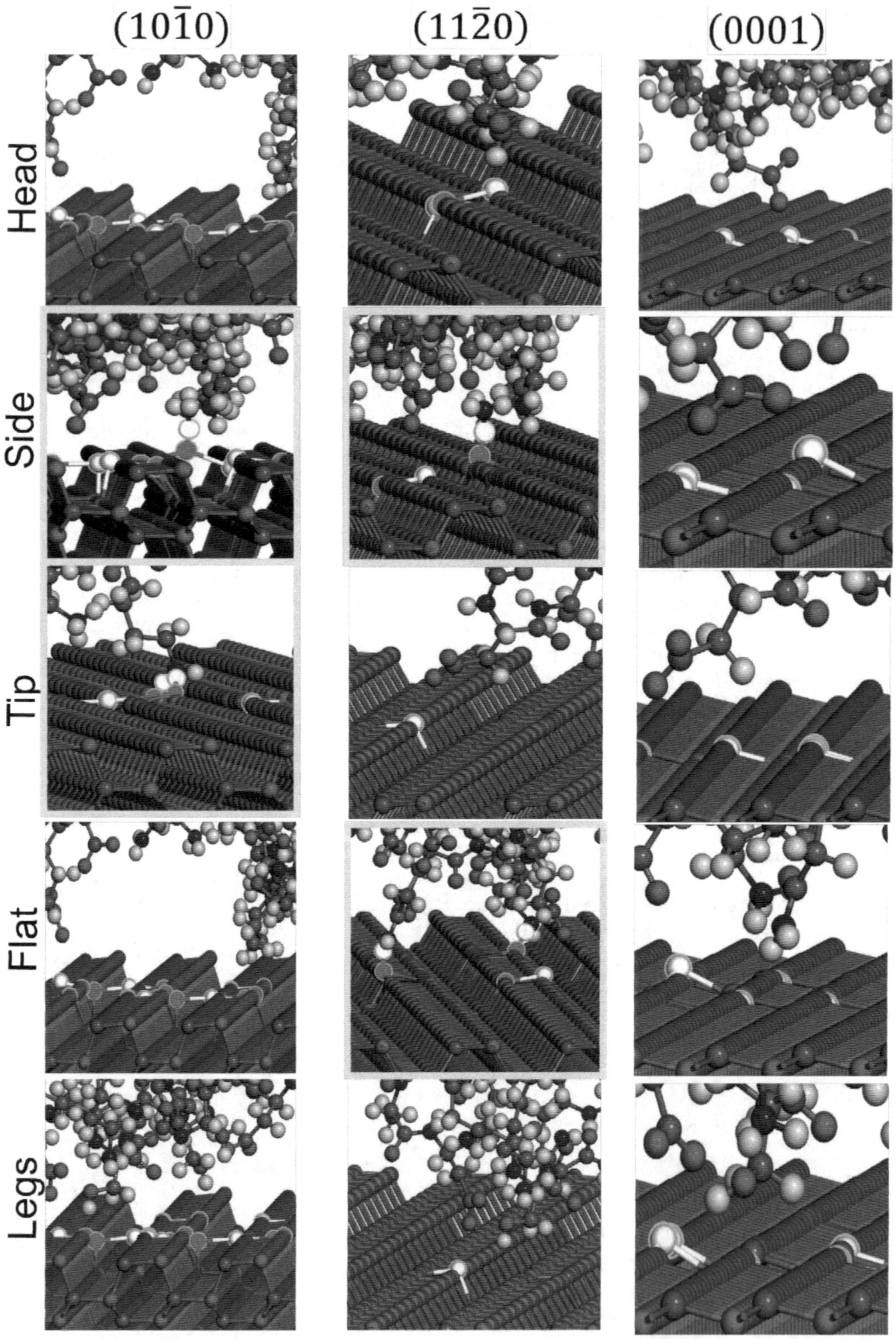

Figure 2. Magnified images of BSA-ZnO interaction sites for different orientations of BSA over (10$\bar{1}$0), (11$\bar{2}$0), and (0001) ZnO planes. The structures are results of force field simulations. Yellow frames denote the configurations in which chemical bonding happens in the calculations.

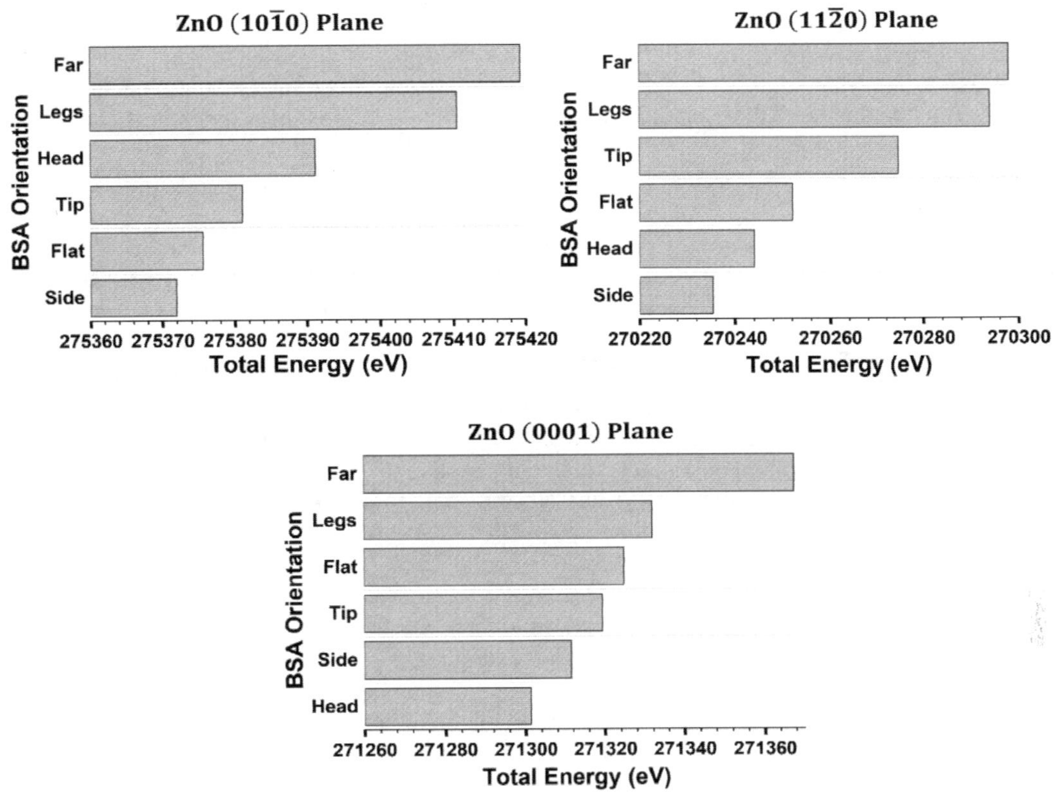

Figure 3. Total energies of BSA+ZnO(10$\bar{1}$0), BSA+ZnO(11$\bar{2}$0), and BSA+ZnO(0001) systems for different orientations of BSA.

Figure 2 summarizes the results of forcefield simulations for BSA molecule over ZnO non-polar and polar planes ((10$\bar{1}$0), (11$\bar{2}$0), and (0001)). Magnified images of the interaction sites between BSA molecule and ZnO planes are illustrated for different orientations of BSA. In all cases, the proximity of BSA molecule to ZnO surface has culminated in mutual attraction, followed by re-arrangement of surface atoms on ZnO slab. In addition to these displacements, chemical bonding of atoms from BSA molecule with atoms on ZnO surface can also be seen at interaction sites in Side and Tip orientations over (10$\bar{1}$0) plane, as well as Side and Flat orientations over (11$\bar{2}$0) plane. In these cases, which are highlighted by yellow frames in figure 2, chemical bonding happens between hydrogen atoms of BSA molecule and surface oxygen atoms of ZnO. For (0001) surface, no chemical bonding has been observed since it contains only Zn atoms compared to nonpolar surfaces which include both Zn and O atoms. It must be mentioned here that more accurate electron-based calculations such as DFT have to be employed to investigate the chemical bonding phenomena and to validate the observed chemical bonds. Nevertheless, the results obtained from the current force field simulations will be helpful for selection of relevant parts of the system for more expensive calculations. Validating the chemical bonding phenomena observed in force field simulations by employing DFT method is going to be done as the next phase of the current study.

To compare different orientations of BSA and determine the most likely one over each ZnO plane, total energy of the system is displayed via bar-graphs in figure 3. Since orientation of BSA molecule is the only varying aspect and other characteristics of the systems remain completely unaffected, total energy of the systems can be observed to deduce the favourable position. According to the total energy graphs in figure 3, tendency of BSA molecule for being adsorbed on ZnO surfaces is noticeably different for different orientations. The head orientation has the lowest energy in the BSA+ZnO(0001) system and thus it is the most likely orientation over ZnO (0001) surface. The side orientation has been observed as the most favourable for BSA+ZnO($10\bar{1}0$) and BSA+ZnO($11\bar{2}0$) systems. It is also worth mentioning that mechanism of interactions is not identical in the most favourable positions over different planes and it depends on the different groups of atoms from the BSA molecule and type of the ZnO plane. Atoms are differently re-arranged, and the adsorption might occur through either physisorption or chemisorption.

4. Conclusion

We employed forcefield simulations to investigate the interactions of BSA molecule with ZnO nonpolar and polar surfaces. The total energy indicates that BSA molecule is likely to adsorb in different orientation on different types of ZnO surfaces. The way how BSA molecule interacts with ZnO surfaces depends on the type of ZnO surface as well as the orientation of the BSA molecule. The computational results demonstrate that the vicinity of BSA molecule with ZnO surfaces results in rearrangement of atoms on ZnO surfaces that lowers the total energy and gives rise to physical binding. Furthermore, in several configurations, chemical bonding between BSA atoms and ZnO surface has also been observed which needs to be validated by more accurate (but more costly) electron-based simulation methods in further studies. This may have important implications for interaction of ZnO with biological environment as well as impact its optical and electronic properties.

Acknowledgements

The work was supported by the Czech Science Foundation (GACR) project 19-02858J.

References

[1] Anand K, Rajamanikandan R, Selva Sharma A, Ilanchelian M, Khan F I, Tiloke C, Katari N K, Boomi P, Balakumar C, Saravanan M, Palanisamy S, Ramesh M, Lai D and Chuturgoon A A 2020 Human serum albumin interaction in silico and anticancer evaluation of pine-gold nanoparticles *Process Biochemistry* **89** 98–109

[2] Wang Y and Ni Y 2013 New insight into protein–nanomaterial interactions with UV-visible spectroscopy and chemometrics: human serum albumin and silver nanoparticles *Analyst* **139** 416–24

[3] Rezek B, Michalíková L, Ukraintsev E, Kromka A and Kalbacova M 2009 Micro-pattern guided adhesion of osteoblasts on diamond surfaces *Sensors (Basel)* **9** 3549–62

[4] Rezek B, Ukraintsev E, Krátká M, Taylor A, Fendrych F and Mandys V 2014 Epithelial cell morphology and adhesion on diamond films deposited and chemically modified by plasma processes *Biointerphases* **9** 031012

[5] Kozak H, Artemenko A, Ukraintsev E, Choukourov A, Rezek B and Kromka A 2019 Infrared absorption spectroscopy of albumin binding with amine-containing plasma polymer coatings on nanoporous diamond surfaces *Langmuir* **35** 13844–52

[6] Ukraintsev E, Rezek B, Kromka A, Broz A and Kalbacova M 2009 Long-term adsorption of fetal bovine serum on H/O-terminated diamond studied in situ by atomic force microscopy *physica status solidi (b)* **246** 2832–5

[7] Kalpana V N and Devi Rajeswari V 2018 A review on green synthesis, biomedical applications, and toxicity studies of ZnO NPs *Bioinorganic Chemistry and Applications* **2018** e3569758

[8] Da Silva E, Kembouche Y, Tegner U, Baun A and Jensen K A 2019 Interaction of biologically relevant proteins with ZnO nanomaterials: A confounding factor for in vitro toxicity endpoints *Toxicology in Vitro* **56** 41–51

[9] Bhaumik A, Shearin A M, Delong R, Wanekaya A and Ghosh K 2014 Probing the interaction at the nano–bio interface using raman spectroscopy: ZnO nanoparticles and adenosine triphosphate biomolecules *J. Phys. Chem. C* **118** 18631–9

[10] Žūkienė R and Snitka V 2015 Zinc oxide nanoparticle and bovine serum albumin interaction and nanoparticles influence on cytotoxicity in vitro *Colloids and Surfaces B: Biointerfaces* **135** 316–23

IOP Conf. Series: Materials Science and Engineering 1050 (2021) 012007 doi:10.1088/1757-899X/1050/1/012007

Effect of ZnO nanoparticle sizes and illumination on growth inhibition of *Escherichia coli* and *Staphylococcus aureus* bacteria in cultivation medium

J Jíra[1], D Rutherford[1], J Mičová[2], Z Remeš[3] and B Rezek[1]

[1] Faculty of Electrical Engineering, Czech Technical University in Prague, Technická 2, 16227 Prague 6, Czech Republic
[2] Institute of Chemistry, Slovak Academy of Sciences, Dúbravská cesta 9, 84538 Bratislava, Slovak Republic
[3] Institute of Physics, Czech Academy of Sciences, Cukrovarnická 10, 16200 Prague 6, Czech Republic

jira@fel.cvut.cz

Abstract. We study antibacterial effects of the zinc oxide nanoparticles (ZnO NPs) in cultivation medium (Mueller-Hinton broth) on Gram-negative (*Escherichia coli*) and Gram-positive (*Staphylococcus aureus*) bacteria models. We compare synthesized ZnO hedgehog NPs and commercial ZnO NPs (50 nm and 20 μm nominal size) in different concentrations (1 mg/mL and 0.1 mg/mL). Results show that *E. coli* are more sensitive to the ZnO presence in the cultivation media than *S. aureus*. We also characterize influence of visible and UV light on the ZnO NP effects.

1. Introduction

Zinc oxide (ZnO) is a material that attracts increasing attention due to its antibacterial abilities, especially as nanoparticles (NPs) [1]. The ZnO particles can be prepared in variety of sizes, shapes and surface modifications. The advantage of nanoscale particle sizes are larger specific surface areas and hence higher surface reactivity compared to larger particles. Small size of NPs enables them specific types of interaction with bacterial cells including penetration into the bacterium structure [2,3]. Special ZnO NP modifications are particles with hedgehog-like shape that showed for instance specific effect on the cultivation of osteosarcoma cells (SAOS-2) [4]. In addition, ZnO is non-toxic to human cells and has a good biocompatibility [5]. This opens the way for usage in the food applications like packaging and bacteria inhibition on food surfaces [6].

In this study we investigate the bacterial response of *E. coli* and *S. aureus* to the ZnO NPs in the Mueller-Hinton (MH) broth. The usage of broth was chosen so that the bacteria and their response would be studied in bacterial natural environment. Three types of ZnO NPs were compared – hedgehog-like ZnO particles and ZnO powder with 50 nm or 20 μm nominal particle sizes. We also studied the effect of illumination on bacterial growth. There was a light source present during the incubation process in broth. The first tested source was a UV lamp and after that we tested also a flatbed visible light source.

2. Materials and methods

Our study involves the use of three different ZnO samples. The samples of nano and micro sizes were obtained from commercial suppliers. The nano sample was purchased from Sigma Aldrich company, declared particle size 50 nm, purity >97%. The micro sample was purchased from the Nanografi nanotechnology company, declared particle size 20 μm, purity >99.9%. The hedgehog-like ZnO needle-clusters were prepared from zinc nitrate hexahydrate ($Zn(NO_3)_2.6H_2O$) by the hydrothermal growth method. More detailed preparation description is in [4].

All ZnO NPs used in the experiments were suspended in demineralized water to prepare the stock solutions with concentration of 2 mg/mL or 0.2 mg/mL. We sterilized these suspensions by autoclave for 30 min. at 120°C. Note that final concentration was 1 mg/mL or 0.1 mg/mL after 1:1 mixing with bacterial suspension. We also tested even lower concentration of the ZnO – 0.03 mg/mL, but this concentration was too low to observe any measurable effects. Typical SEM morphology of the employed ZnO nanoparticles is shown in figure 1. The samples were drop-casted onto silicon substrates. The images show us that the 50 nm NPs and 20μm NPs have similar crystalline structure, but the crystals are different in size. The structure of hedgehog-like ZnO shows predominantly clusters of needle-like shapes with sizes of few micrometres.

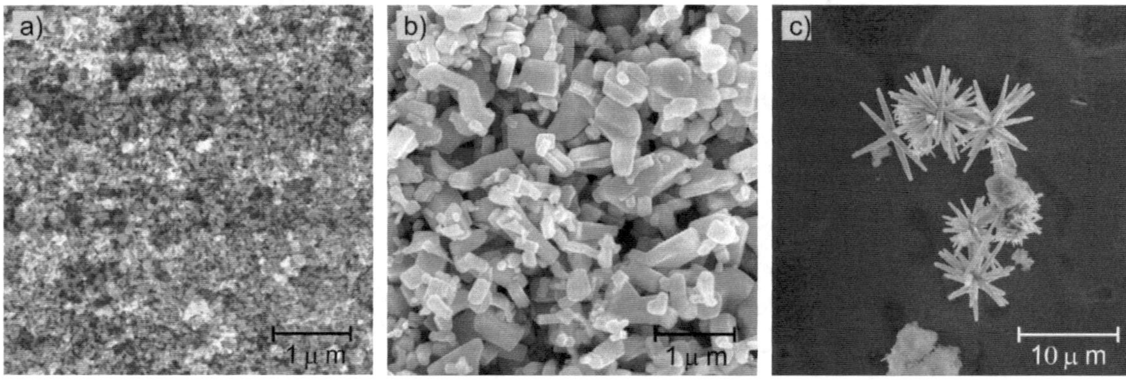

Figure 1. SEM morphology of the employed ZnO samples: ZnO 50 nm (a), ZnO 20 μm (b) and the ZnO hedgehogs (c).

Dynamic light scattering (DLS) and zeta potential were measured on colloidal suspensions of ZnO particles in the MH broth at 25 °C using Nano-ZS (Malvern, UK) equipped with a He-Ne laser. The ZnO samples were not ultrasonicated to prevent damage of microstructures. The samples were mixed with the MH broth by vortex to achieve the final concentration of 1 mg/mL.

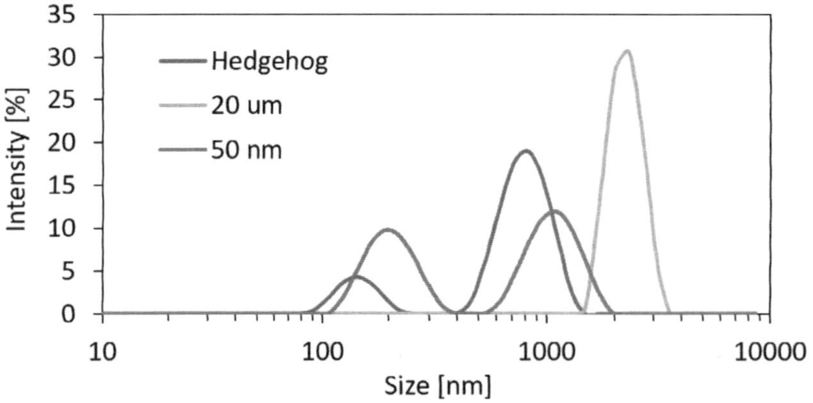

Figure 2. DLS measurement of ZnO samples mixed in MH broth.

Figure 2 shows that the size of nominally 20 μm ZnO sample is actually in units of μm while the size of nominally 50 nm ZnO sample is actually between 100-300 nm with some clusters of size around 1 μm. The ZnO hedgehogs have micron size with some fragments of 100-200 nm size. Another important parameter for interaction with cells is zeta potential. The zeta potential is important for the stability of nanoparticles in suspension. It depends on the surface charge and is also the major factor in the initial adsorption of nanoparticles onto the cell membrane. After adsorption, the endocytotic uptake rate depends on the particle size. It was measured in the same DLS apparatus using the same ZnO samples in MH broth. The resulting values are listed in the table 1. All ZnO samples exhibit negative zeta potential. Its value is relatively small, between -12 to -18 mV, still the colloidal solutions are reasonably stable as shown also by the DLS measurements.

Table 1. Zeta potentials of ZnO samples in MH broth.

Sample	Hedgehog ZnO	ZnO 50 nm	ZnO 20μm
Zeta potential [mV]	-11.7	-17.5	-15.5

The bacterial solutions were prepared in the following manner. We spread 1 ml of concentrated *E. coli* or *S. aureus* solution from the freezer after 15 min melting on the 9 cm Petri dish with the MH agar and let it grow overnight in the thermostat at the temperature 37 °C. Then all the bacteria from the Petri dish were wiped off and put into 5 ml of the MH broth. This basic concentration of the bacteria was considered as 10^0. This initial solution was diluted by the broth in 1:10 ratio several times until we achieved its turbidity around 0.5 MFU (MacFarland Units). The closest value to this turbidity had always a sample with relative concentration of 10^{-3}, which roughly corresponds to the colony forming unit (CFU) concentration of 10^7 per mL. This concentration was used for our experiments.

Each experiment was performed in four test tubes, where one was a reference containing a mixture of bacterial solution and demi water (3+3 mL) and three other tubes contained a mixture of bacterial solution and solution with one of the three examined ZnO samples (again 3+3 mL). Since we examined *E. coli* and *S. aureus*, and we used two different ZnO concentrations, we did four separate experiments.

The tubes were put into the shaker inside the incubator keeping 37 °C. We took samples from each tube at 0, 3, 5, 7, 24 hours to obtain a growth curve and to catch both exponential and stationary phase of the growth. Each sample was diluted to the concentration suitable for further Petri dish cultivation. The suitable concentration wad determined by a preliminary experiment using 1:10 dilution steps. Each diluted solution was spread in the amount of 1 mL on 9 cm Petri dish with the MH agar in triplicates (three dishes per each sample). The dishes were put into the incubator at 37 °C and let grow for 24 hours. After that we counted the number of bacterial colonies on Petri dishes and we calculated CFU concentrations for each sample. Finally, we calculated the R-value, the logarithmic ratio between the CFU number of reference and the CFU number of evaluated sample.

As an additional experiment we also measured the bacterial response to the ZnO NPs with a light source present during the incubation of the tubes. The first source was a UV lamp Philips TUV 15W/G15 T8, which had the spectral maximum at the wavelength of 250 nm. Then we tested a visible light source represented by a light panel Reflecta L230 generating fluorescent light. For these experiments we used two pairs of reference and ZnO samples, when one pair of tubes was packed in aluminium foil and the second pair was exposed to the light. Only one ZnO sample was chosen for simplicity – the hedgehog. The distance between the tubes and light source was around 15 cm in both cases.

3. Results and discussion

The main results of our experiments are depicted in figure 3. There are growth curves for both types of used bacteria in both used concentrations of the three types of ZnO particles. The results show that the *E. coli* bacteria are more sensitive to higher concentrations (1 mg/mL) of the ZnO structures than *S. aureus*. The best antibacterial activity against the *E. coli* was observed for the hedgehog NPs, where we achieved the highest ratio of the reference CFU number compared to measured sample, which was six orders here in the stationary phase. The response of *S. Aureus* to higher ZnO concentration was similar for all types of ZnO structures, resulting in a four order difference between the CFU number of reference and ZnO samples.

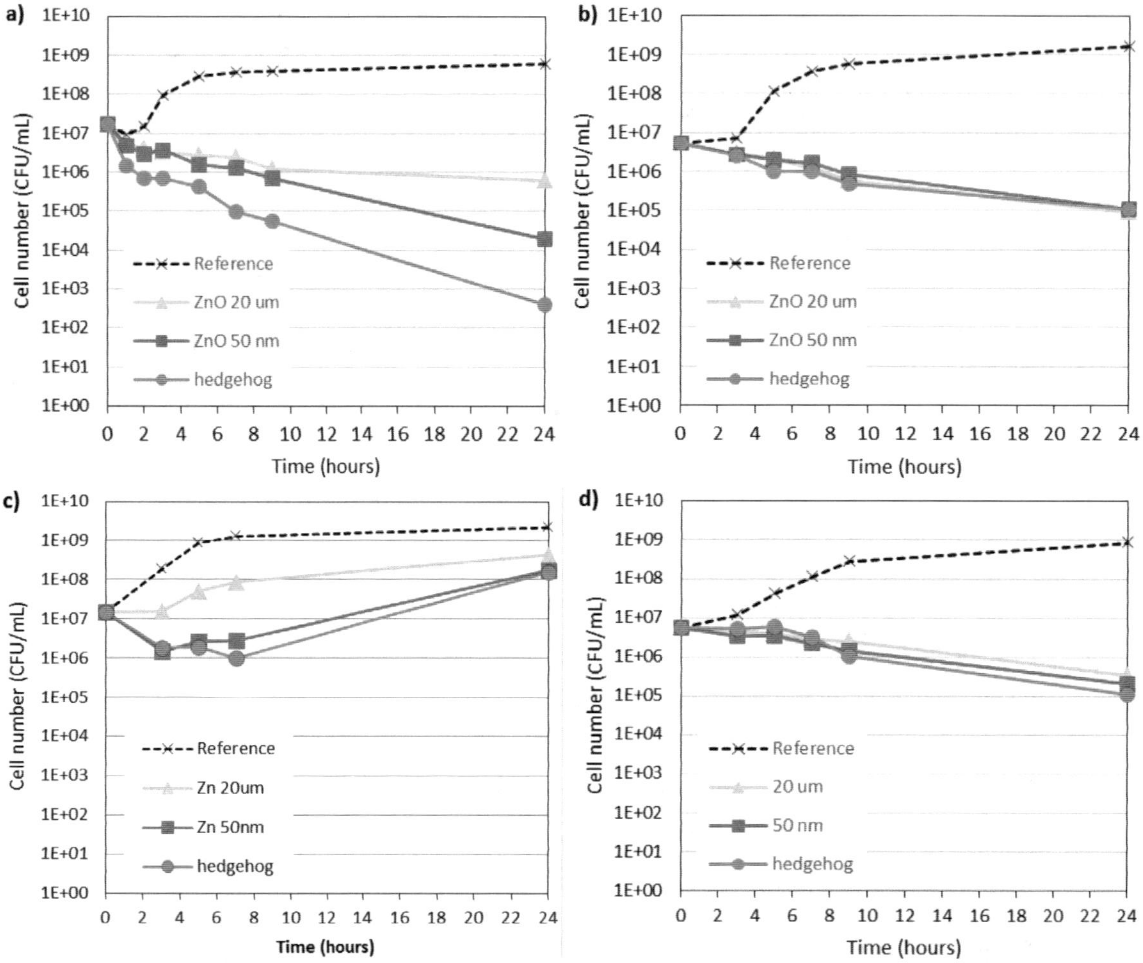

Figure 3. Growth curves of the *E. coli* with 1 mg/mL ZnO (a), *S. aureus* with 1 mg/mL ZnO (b), *E. coli* with 0.1 mg/mL ZnO (c) and *S. aureus* with 0.1 mg/mL ZnO (d).

The response to lower concentration (0.1 mg/mL) of the ZnO particles was different. While *S. aureus* had similar response resulting in almost four orders ratio compared to the reference, the effect on *E. coli* was lower, only up to 2.5 orders during the exponential phase and only one order in the stationary phase.

The results were also interpreted in the form or R-value, where the R-value was calculated according to the formula $R = \log(C/E)$, where C is the CFU number of the reference sample and E is the CFU number of the measured sample. The R values for the *E. coli* and *S. aureus* measurements for both concentrations of ZnO are summarized in the bar graphs in figure 4. The sampling times for the bar graphs were selected so that they characterize both exponential phase of the growth curve (5 hours) and stationary phase of the curve (24 hours).

We can see from the bar graphs that the response of *S. aureus* is more uniform in both exponential and stationary phase, achieving the R-value up to the value of 4 for the high ZnO concentration. The response of the *E. coli* is more ZnO type dependent where the hedgehog shows the best antibacterial results with the R-value up to the 6 for higher ZnO concentration. There is also notable that lower ZnO concentration against *E. coli* is the only configuration, where the R-value drops from the exponential phase towards the stationary one and the bacteria are able to partially recover.

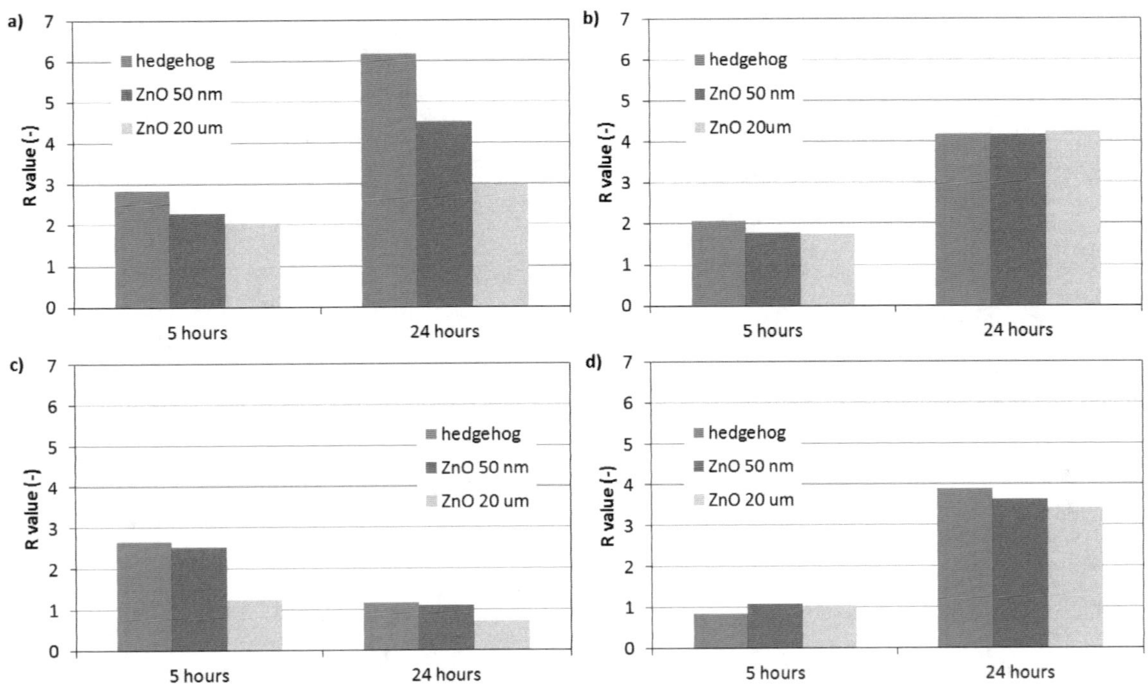

Figure 4. R-value graphs of the *E. coli* with 1 mg/mL ZnO (a), *S. aureus* with 1 mg/mL ZnO (b), *E. coli* with 0.1 mg/mL ZnO (c), and *S. aureus* with 0.1 mg/mL ZnO (d).

The experiments with visible and UV light sources provided different results. We performed the experiments only with the hedgehog ZnO in concentration of 1 mg/mL, because they showed the best antibacterial effect. The visible light had no significant effect on bacterial growth, which is demonstrated by the growth curves in figure 5.

One can also notice good reproducibility of the experiments. Contrary to that, the UV illumination resulted in a very strong antibacterial effect. All bacteria in solutions that were not packed in the aluminium foil were killed, both in the ZnO and reference sample, so we could not distinguish between the effect of ZnO itself and additional UV light. The data is thus not plotted.

The antibacterial mechanisms of the ZnO particles are still a subject of discussion. There are several proposed mechanisms of such ability [1,3]. The first mechanism is the generation of reactive oxygen species (ROS), which can destroy cellular components like lipids, proteins etc. The generation of ROS is mostly photocatalytic. Our experiments were performed mostly in the dark. Illumination by visible light had also no noticeable effect. The ROS mechanism will be thus minor in our case.

The second mechanism is the release of zinc ions Zn^{2+} that can affect the amino acid metabolism and enzyme system disruption. The ability of releasing zinc ions is surface dependent, hence the small ZnO NPs, that have larger surface area to volume ratio than larger particles, are supposed to have better antibacterial ability due to this mechanism. Comparison of our results of 50 nm and 20 μm particles seems to confirm that. However, the ZnO hedgehogs have smaller surface area yet superior bactericidal effect on *E. coli*.

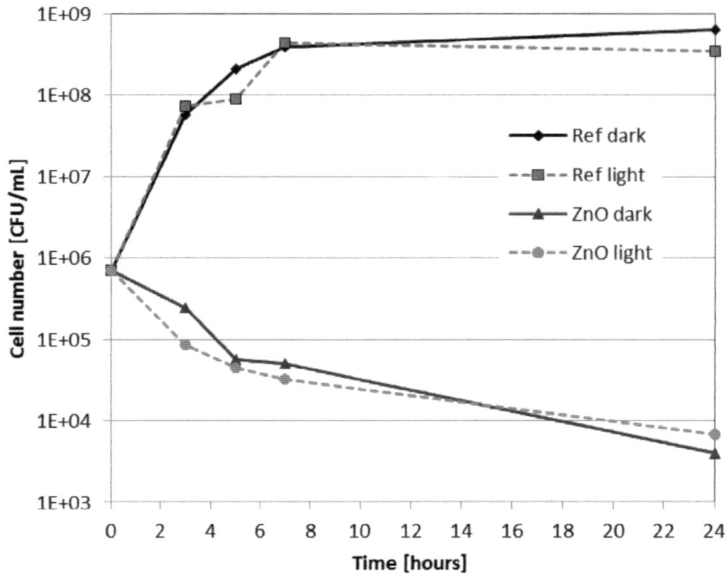

Figure 5. Growth curves for the additional visible light experiment. Dark means a sample packed in aluminium foil, light means a sample exposed to the visible light.

The third possible mechanism is electrostatic interaction. The surface of bacteria is negatively charged. The electrostatic forces can then lead to create an envelope of the ZnO particles around bacteria which could result in mechanical disruption of bacterial membranes. Similar effect has been observed for nanodiamond particles [7]. Smaller ZnO particles have smaller size and larger number per weight, thus they have better chance to create such envelope compared to large ones. This would again agree with the data for 50 nm and 20 μm particles. However, the ZnO hedgehogs have superior effect although they are large and less concentrated. Moreover, zeta potentials of all ZnO materials are negative and quite low (see table 1). The electrostatic interaction thus cannot make a predominant effect.

The fourth possible mechanism is a mechanical damage of bacterial cell walls caused by specific shape of the ZnO NPs [8]. The hedgehog needles have such specific shape, with numerous spikes in all directions. Relatively large size (micrometers) provides them also with higher kinetic energy during bacteria cultivation in the shaker. The ZnO hedgehogs thus may exert their antibacterial effect mainly by mechanical damage of bacteria. This is pronounced especially against the Gram-negative bacteria like *E. coli* that have thinner peptidoglycan layer (2-7 nm [9]) compared to Gram-positive *S. aureus* with thick peptidoglycan layer (20-30 nm [10]). This effect is observable when using higher concentration of ZnO structures, namely 1 mg/mL. As for lower concentration of 0.1 mg/mL the effect is much less pronounced for all types of ZnO, the mechanical disruption seems in all cases to be the dominant effect compared to other factors described above.

4. Conclusions

We found that the ZnO particles, when used in higher concentration of 1 mg/mL, had significant antibacterial effect suppressing the bacterial growth in MH broth by up to six orders of magnitude compared to the reference after 24 hours. Lower concentrations of 0.1 mg/mL had significant effect only on the *S. aureus* while the *E. coli* indicated only inhibited growth finishing one order down compared to the reference after 24 hours. Smaller 50 nm ZnO particles had better antibacterial effect against the *E. coli* compared to larger 20 μm particles. However, ZnO particles with hedgehog structure had the best antibacterial effect against the *E. coli* compared to the particles with crystallite structure. Visible light did not affect the bacteria growth or the ZnO antibacterial activity. The mechanical disruption thus seems to be the dominant effect compared to other possible antibacterial mechanisms. This is in agreement with the observation that the antibacterial effect against Gram-positive *S. aureus* with robust membrane was similar for all employed types of ZnO particles.

Acknowledgements
Raisa Yatskiv is kindly acknowledged for the help with SEM imaging. The work has been financially supported by the project GACR 19-02858J, by the project from Operational Programme Research, Development and Education financed by the European Structural and Investment Funds and the Czech Ministry of Education, Youth and Sports (CZ.02.1.01/0.0/0.0/16_019/0000760 – Solid21) and by the Scientific Grant Agency (VEGA, Slovak Academy of Sciences, Grant No 2/0157/20). We also acknowledge the use of CzechNanoLab Research Infrastructure supported by the MEYS CR (LM2018110).

References

[1] Sirelkhatim A, Mahmud S, Seeni A, Kaus N H M, Ann L C, Bakhori S K M, Hasan H and Mohamad D 2015 Review on zinc oxide nanoparticles: Antibacterial activity and toxicity mechanism *Nano-Micro Lett.* **7** 219–42
[2] Rasmussen J W, Martinez E, Louka P and Wingett D G 2010 Zinc oxide nanoparticles for selective destruction of tumor cells and potential for drug delivery applications *Expert Opin Drug Deliv* **7** 1063–77
[3] Mishra P K, Mishra H, Ekielski A, Talegaonkar S and Vaidya B 2017 Zinc oxide nanoparticles: A promising nanomaterial for biomedical applications *Drug Discovery Today* **22** 1825–34
[4] Neykova N, Brož A, Remeš Z, Hruška K, Kalbáčová M, Kromka A and Vaněček M 2012 ZnO hedgehog-like structures for control cell cultivation *Applied Surface Science* **258** 3485–9
[5] Padmavathy N and Vijayaraghavan R 2008 Enhanced bioactivity of ZnO nanoparticles - an antimicrobial study *Sci Technol Adv Mater* **9** 035004
[6] Chaudhry Q and Castle L 2011 Food applications of nanotechnologies: An overview of opportunities and challenges for developing countries *Trends in Food Science & Technology* **22** 595–603
[7] Jira J, Rezek B, Kriha V, Artemenko A, Matolínová I, Skakalova V, Stenclova P and Kromka A 2018 Inhibition of E. coli growth by nanodiamond and graphene oxide enhanced by Luria-Bertani medium *Nanomaterials* **8** 140
[8] Zanni E, Bruni E, Chandraiahgari C R, De Bellis G, Santangelo M G, Leone M, Bregnocchi A, Mancini P, Sarto M S and Uccelletti D 2017 Evaluation of the antibacterial power and biocompatibility of zinc oxide nanorods decorated graphene nanoplatelets: New perspectives for antibiodeteriorative approaches *J Nanobiotechnol* **15** 57
[9] Gumbart J C, Beeby M, Jensen G J and Roux B 2014 Escherichia coli peptidoglycan structure and mechanics as predicted by atomic-scale simulations *PLoS Comput Biol* **10** e1003475
[10] Sharif S, Singh M, Kim S J and Schaefer J 2009 Staphylococcus aureus peptidoglycan tertiary structure from carbon-13 spin diffusion *J Am Chem Soc* **131** 7023–30

IOP Conf. Series: Materials Science and Engineering 1050 (2021) 012008 doi:10.1088/1757-899X/1050/1/012008

Origami-kirigami approach to materials structures modelling

I Markechová and H Stúpalová

Slovak University of Technology in Bratislava, Faculty of Materials Science and Technology in Trnava, Ulica Jána Bottu 2781/25, 917 24 Trnava, Slovak Republic

iveta.markechova@stuba.sk

Abstract. Additive manufacturing and origami/kirigami principles fit together like a coffee cup and saucer. Recently derived translational surface with an elliptic curve (not an ellipse), as the governing one, led to constructing origami-like object. Set of eight surfaces, each with a square-shaped orthogonal projection, turned out to be enchantingly similar to heaven-and-hell-origami. In presented work engaged starting translational surface (STS) is originally subjected to the selected linear and nonlinear transformations, in two ways. The first one is dedicated to spacial placement of STS replicas. Shape STS variations represent the second approach. Square-shape of STS orthogonal projection is preserved during all the mappings, it does not change. On the other hand, especially in this point, the similarity with origami/kirigami structures is broken; generally, square-shaped sheet of paper changes via folding. In our paper, preserved orthogonal projection offers suitable opportunities to illustrate a structure of the hypothetical material. Explicit and parametric equations of geometrical elements, transformational matrices and MATLAB application MuPAD serve as the useful mathematical and computational tools for formal and graphical representation of the modeled structures.

1. Introduction

Grey eminence of what follows is an elliptic curve (EC). Not an ellipse. Nor a part of an ellipse. But an elliptic curve. Its significant and typical field of application is ECC – elliptic curve cryptography. Not a materials structures modelling.

In the last years, new technological alternatives have initialized an opening the door to the advent of novelty materials. On the one hand novelty from the viewpoint of structure/geometry. And simultaneously from physical properties perspective. References [1-13] represent multi-taste spectrum.

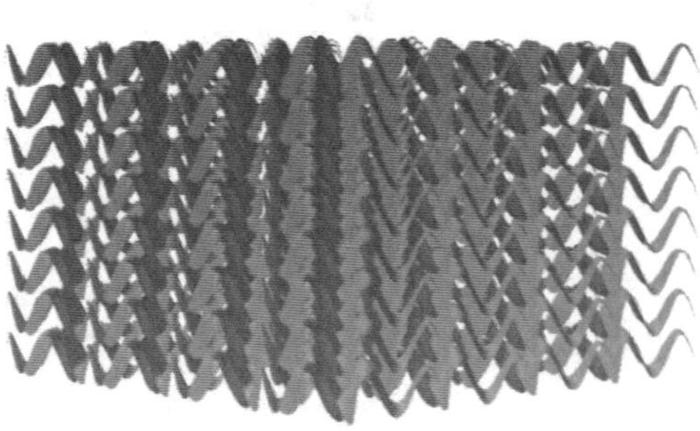

Figure 1. 3D geometrical structure of hypothetical material. An elliptic curve is a governing key curve of translational surface element.

We start from findings in [14] and [15] and their specific relation to origami. We offer an original, simple approach to modelling of hypothetical materials structures as can be seen in Figure 1. There is a future challenge to look into the role of geometrical properties of EC in physical aspects of material structures described herein. By then: brief creating-material-architecture-tour.

The aims of a next section are twofold: a) to sketch a key translational surface (TS) element with EC as a governing one [14-15] and b) to give a mathematically formal and a graphical illustration of a structure with solely linear mappings of TS as well as with TS-shape conservation. The changes of the TS-form as a consequence of nonlinear transformations are discussed in section 3. Selected nonlinear mappings do not modify squared orthogonal projection of TS; thank to this, elements arranged in a row have continuous projection, nevertheless transformed TS elements are of 3D shapes, more or less isolated and similar to kirigami structures. Replicas of transformed TS are arranged to layer, eventually layers are posed on top of each other.

2. Elliptic curve in a starting translational surface element and preserving-the-object-shape linear transformations

Hoping in clarity of interpretation 😉 we will continue by naming verbal, formal and graphical (FGV) representations of the objects. Here they are, one after the other:

- implicit equations of the elliptic curve, parameters a, b are fixed,
- explicit equations of the elliptic curve, part for a positive Cartesian coordinate y,
- parametric equations of translational surface; EC is engaged to the position of each of both governing curves; dividing by 2 is for to obtain less slope of the curve;

$$y^2 = x^3 + \boldsymbol{a}x + \boldsymbol{b},\ (a, b) = (1, 0) \tag{1}$$

$$y(x) = \sqrt{x^3 + x} \tag{2}$$

$$\boldsymbol{r}(t, u) = (x(t, u), y(t, u), z(t, u)) = \left(t, u, \left(\sqrt{t^3 + t} + \sqrt{u^3 + u}\right)/2\right) \tag{3}$$

where t, u vary from 0 to 1.

Translational surface (Figure 2) serves as the basic element in the geometrical sets. In the following, firstly symmetry with respect to a horizontal plane is done. Plane intersects starting TS in the middle of its height. Duet of starting TS and a planarly symmetrical one are shown in Figure 3 from various viewpoint. TS duets are spatially arranged to the layer (Figure 4).

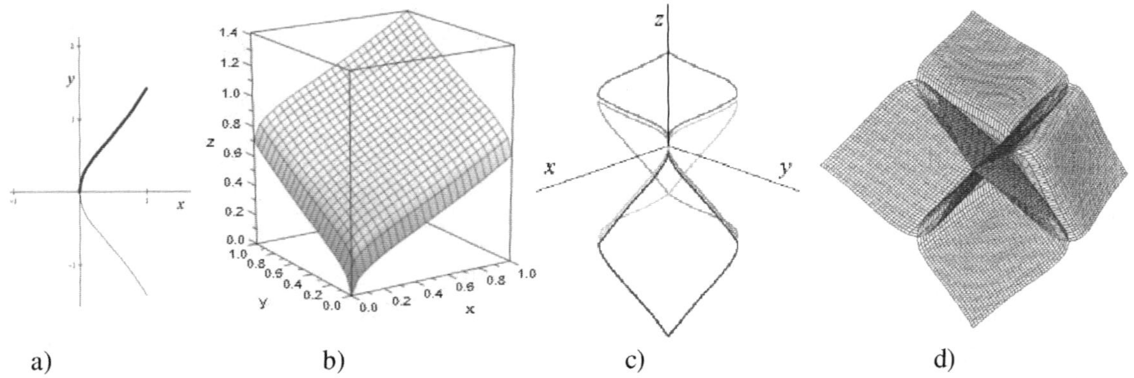

a) b) c) d)

Figure 2. There are step by step a) 2D elliptic curve (EC), b) 3D translational surface: look at the borders of the surface – these are ECs, c) quartet of TSs, d) octet of TSs, which is similar to heaven-and-hell-origami-like object [14-15].

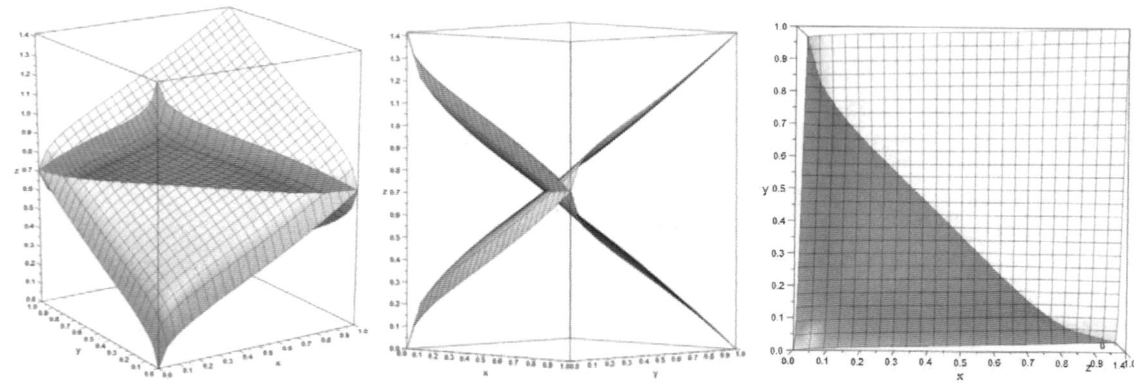

a) b) c)

Figure 3. Duet of translational surface and its planarly symmetrical one in a three different positions. Figure 3c) represents a view from the above.

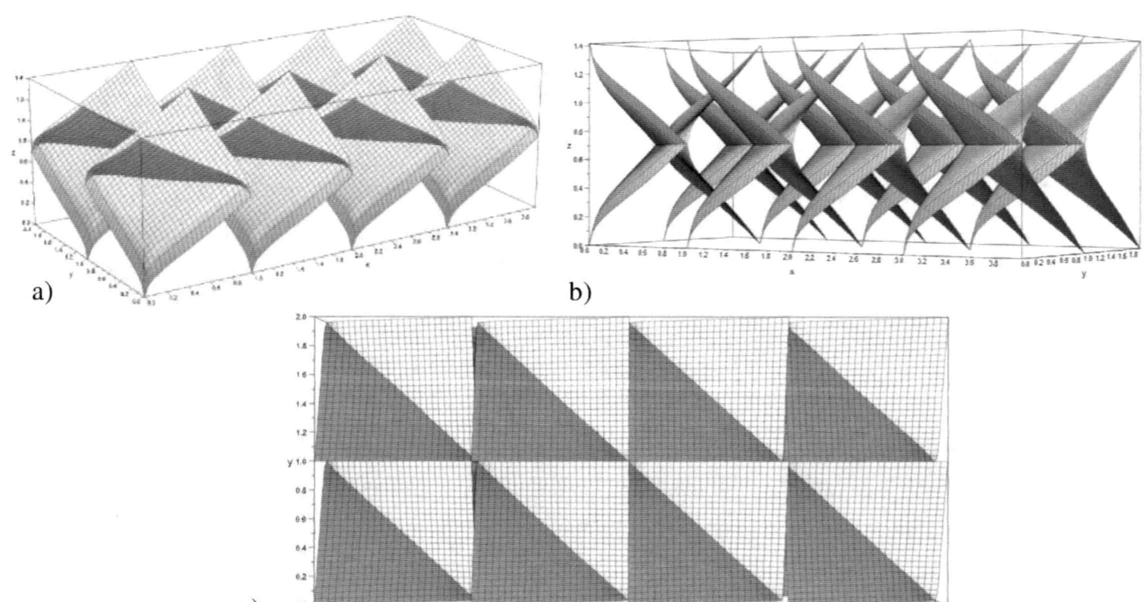

a) b) c)

Figure 4. Three different views on the doublet-layer.

3. Nonlinear transformations

Sine-function was chosen to the experiments with nonlinear transformations of starting TS. In parametrical equations, x and y coordinates are without changes. Nonlinearity is placed to z-coordinate in two different ways:

- original z-coordinate is multiplied by sine,
- z-coordinate becomes an argument of a sine.

Here are the corresponding analytical expressions:

$$\boldsymbol{r}^*(t, u) = (x(t, u), y(t, u), z^*(t, u)) = \left(t, u, f^*(t,u)\left(\sqrt{t^3+t}+\sqrt{u^3+u}\right)/2\right) \tag{4}$$

$$f^{*}(t,\,u) = A\sin(Btu),\quad (A,\,B) = (1/10,\,24) \tag{5}$$

$$\boldsymbol{r}^{**}(t,\,u) = (x(t,\,u),\,y(t,\,u),\,z^{**}(t,\,u)) = (x(t,\,u),\,y(t,\,u),\,A\sin(Bz(t,u)) =$$

$$= \left(t,\,u,\,A\sin\frac{B\left(\sqrt{t^3+t}+\sqrt{u^3+u}\right)}{2}\right) \tag{6}$$

$$z^{**}(t,\,u) = A\sin(Bz(t,u)),\quad (A,\,B) = (1/6,\,16), \tag{7}$$

where t, u vary from 0 to 1, in the same way as in the case of the previous linear mappings. Finally, graphical representation is shown in Figure 5 and Figure 6.

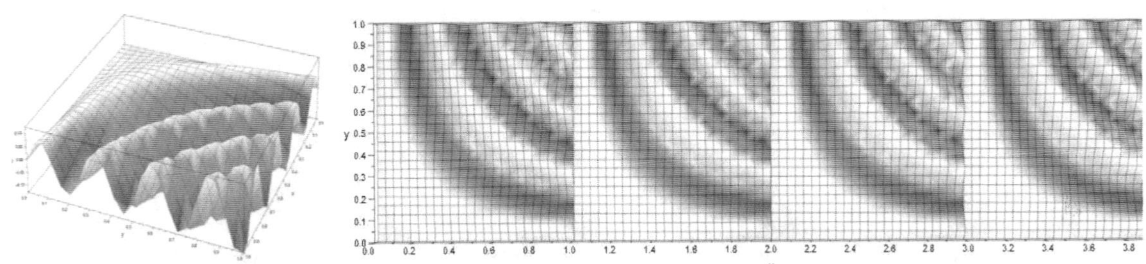

a) b)

Figure 5. Two different views on the nonlinearly transformed TS and the series of its replicas.

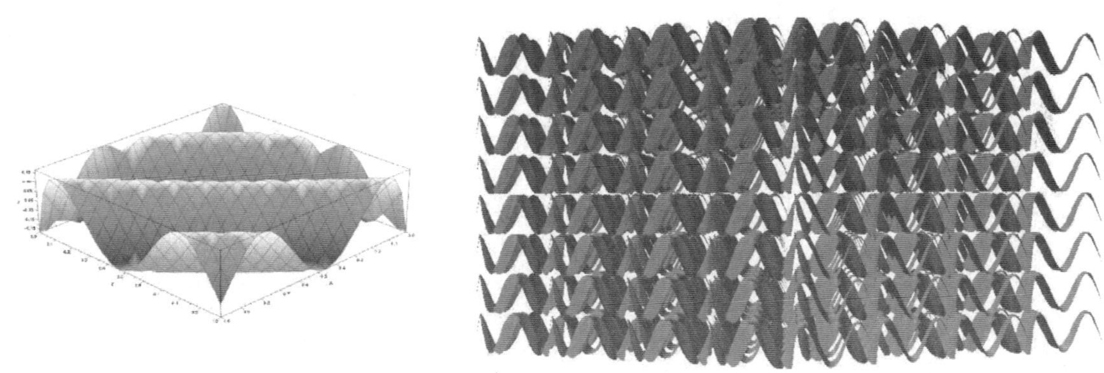

a) b)

Figure 6. The second nonlinear transformation. One shape-changed TS element (a) and the set of layers from these nonlinearly transformed surfaces (b).

4. Conclusion and a future work

It is a question of a future work to develop existing knowledge at these promising directions:

a) variations in the very beginning – at the heart of the matter: engaging the **elliptic curves** of the different $(a,\,b)$ parameters and consequently of the various differential-geometry-features,

b) **geometry** variations in a shape and a position of the STS, i. e. involvement other choices from infinity menu of linear and nonlinear transformations,

c) research of **physical** properties of simulated or real structures,

d) investigation/**control** in b) and c) issues with respect to a possible exploiting of a **functional gradient**, taking into account existing EC analytical representation via explicit, implicit and/or parametric functions.

Appendix

An incredible coincidence from the article preparation phase – snapshot from the copy center: kirigami-like pattern on the lady dress (Figure 7). This structure is similar to the one on the Figure 8 [2]; original figure is here rotated ninety degrees.

Figure 7. Pattern on the fabric.

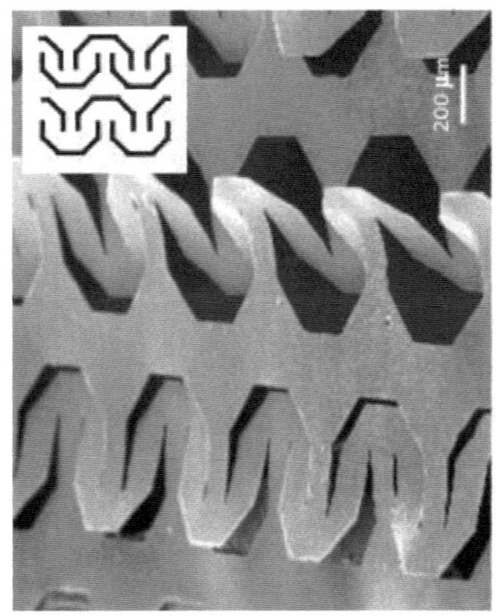

Figure 8. Microscale kirigami pattern [2].

Acknowledgments

This publication is the outcome of the "UNIVERSITY SCIENTIFIC PARK: CAMPUS MTF STU – CAMBO"project implementation (ITMS: 26220220179) supported by the Research & Development Operational Program funded by the EFRR. This contribution was also funded by the institutional project 1609 04/2018-12/2019 of "Relation elliptic curve - ellipse from the viewpoint of selected differential equations and geometric design of virtual model of robogami". The first author is especially grateful to Marek Rariga for his creative interest in the theme within a bachelor thesis framework.

References

[1] Chen S et al. 2020 Kirigami/origami: unfolding the new regime of advanced 3D micro-fabrication/nanofabrication with "folding" *Light Sci. Appl.* **9** 1 1-19.

[2] Shyu T C et al 2015 A kirigami approach to engineering elasticity in nanocomposites through patterned defects *Nat. Mater.* **14** 8 785-789.

[3] Ning X at al 2018 Assembly of advanced materials into 3D functional structures by methods inspired by origami and kirigami: a review *Adv. Mater. Interfaces* **5** 13 1800284.

[4] Rus D and Tolley M T 2018 Design, fabrication and control of origami robots. *Nat. Rev. Mater.* **3** 6 101-112.

[5] Hawkes E et al 2010 Programmable matter by folding *Proc. Natl. Acad. Sci.* **107** 28 12441-5.

[6] Moshe M et al 2019 Nonlinear mechanics of thin frames *Phys. Rev. E* **99** 1 013002.

[7] Sussman D M et al 2015 Algorithmic lattice kirigami: A route to pluripotent materials *Proc. Natl. Acad. Sci.* **112** 24 7449-7453.

[8] Callens S J P, Tumer N and Zadpoor A A 2019 Hyperbolic origami-inspired folding of triply periodic minimal surface structures *Appl. Mater. Today* **15** 453-461.

[9] Al-ketan O et al 2020 Functionally graded and multi-morphology sheet TPMS lattices: Design, manufacturing, and mechanical properties *J. Mech. Behav. Biomed. Mater.* **102** 103520.

[10] Jenett B et al 2020 Discretely assembled mechanical metamaterials *Sci. Adv.* **6** 47 eabc9943.

[11] Shigemune H et al 2020 Programming Stepwise Motility into a Sheet of Paper Using Inkjet Printing *Adv. Intell. Syst.* 2000153.

[12] Hu F et al 2020 Origami spring–inspired metamaterials and robots: An attempt at fully programmable robotics *Sci. Prog.* **103** 3 0036850420946162.

[13] Medina H and Kohler R 2020 Novel rosette-like formations with nanothick petals observed on acid-etched titanium *Engineering Reports* **2** 10 e12247.

[14] Markechová I and Hamplová L. Čipková 2016 Origami robot model via an elliptic curve? In: *2016 IEEE 14th International Symposium on Intelligent Systems and Informatics (SISY)*. IEEE 167-170.

[15] Markechová I and Stúpalová H 2018 Functionally Graded Virtual Robogami: A Geometry Approach In: *2018 10th International Conference on Electronics, Computers and Artificial Intelligence (ECAI)*. IEEE 1-4.

Dynamic Analysis of Cone Waveguide for Ultrasonic Assisted Technological Processes

M Naď, Š Šimon and P Bucha

Slovak University of Technology in Bratislava, Faculty of Materials Science and Technology in Trnava, Ulica Jána Bottu 2781/25, 917 24 Trnava, Slovak Republic

milan.nad@stuba.sk

Abstract. Research and development of new materials (ceramics and composites) with advanced mechanical properties also leads to the development and application of new so-called hybrid technological processes, which are based on a combination of conventional technological processes with ultrasonic tool vibrations. The fundamental condition for the optimal realization of technological process is the vibration of ultrasonic waveguide in the resonant state. Therefore, the ultrasonic waveguide must have the required modal properties, which depend on the distribution of its mass and stiffness properties. Design and analysis of dynamic properties of ultrasonic cone waveguide with adaptive modification of modal properties is presented in this paper. Modification of modal properties is performed using an embedded core, which changes the distribution of spatial properties of the waveguide structure. The effect of geometric parameters and material properties of the inserted core on the ultrasonic waveguide modal properties (mode shapes, natural frequency, amplification factor) is presented.

1. Introduction

The high requirements for material properties that are currently used result in the development of qualitatively new materials and material structures. These new advanced materials (ceramics, composites, etc.) are generally characterized by improved physical and mechanical properties. This situation, of course, leads to relatively large problems in their technological processing. One of the ways to perform the required technological process and achieve a quality of required processing of these materials are technological processes using the so-called hybrid technology processes [3-4]. Hybrid technological processes are usually based on a combination of conventional technological processes with ultrasonic vibration. The transmission of vibrations to the place of realization of the technological process (for example in the case of turning to the cutting edge of a turning tool) is performed in an electromechanical excitation system [1]. The mechanical part of electromechanical excitation system (Figure 1) consists of an ultrasonic generator, electromechanical transducer, booster and waveguide (ultrasonic horn or sonotrode). In order to achieve the required parameters of the hybrid technological process, a resonant state must be achieved [2] in the whole mechanical subsystem of the electromechanical excitation system. The ultrasonic waveguide is the most important part of the whole excitation system, which have to be in the resonant state because it performs the transmission of the oscillation amplitude directly into the performed technological process. However, due to the changes that occur during the execution of the technological process, the "detuning" of the resonant state of the excitation system occurs and the quality of the performed technological process is unsatisfactory. To preform of required technological process, it is necessary to change the parameters of the excitation system so that the mechanical part of the excitation system will be still tuned in the resonant state [2]. Design and analysis of cone ultrasonic waveguide with adaptive change of modal properties performed by embedding of reinforcing core is solved in this paper.

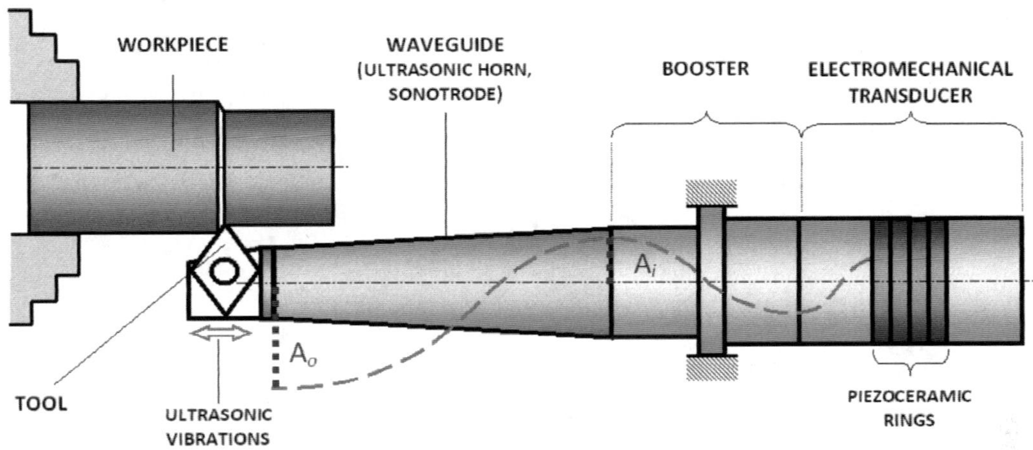

Figure 1. Model mechanical part of ultrasonic excitation systems.

2. Formulation of the problem

The mathematical formulation of the longitudinal vibration of a general waveguide is derived from the waveguide model shown in Figure 2.

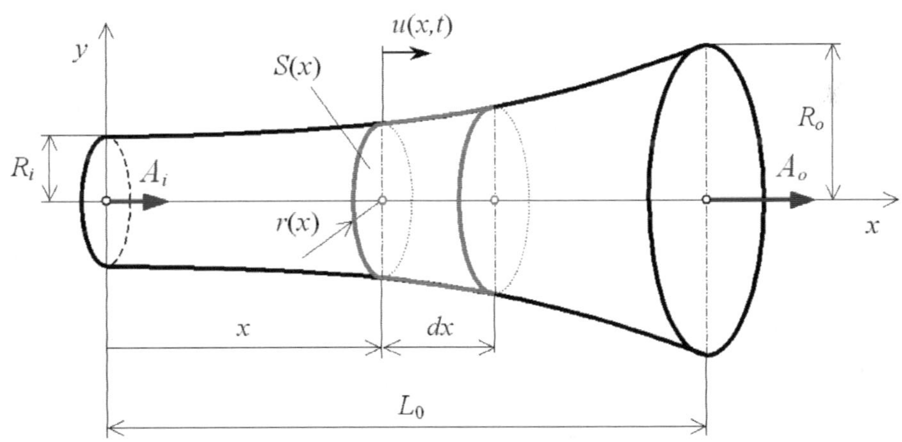

Figure 2. Ultrasonic waveguide design for general geometrical shape.

The partial differential equation (PDE) describing the free longitudinal vibration of general waveguide structure [5-6] is defined in the following form

$$\rho S(x)dx\frac{\partial^2 u(x,t)}{\partial t^2}-\frac{\partial}{\partial x}\left(ES(x)\frac{\partial u(x,t)}{\partial x}\right)dx=0 \tag{1}$$

where x - coordinate in longitudinal direction,

L_0 - waveguide length,

$u(x, t)$ - longitudinal displacement of cross section in x direction,

$S(x)$ - cross section in x position,
E - Young modulus,
ρ - density of waveguide material,

After the introduction and substitution of dimensionless quantities

- dimensionless coordinate $\xi = \frac{x}{L_0}$, (2)
- dimensionless displacement $\overline{u}(\xi,t) = \frac{u(x,t)}{L_0}$ (3)

the equation (1) has the form

$$\frac{\partial^2 \overline{u}(\xi,t)}{\partial t^2} - \frac{c_p^2}{L_0^2} \frac{1}{f_s(\xi)} \frac{\partial}{\partial \xi}\left(f_s(\xi) \frac{\partial \overline{u}(\xi,t)}{\partial \xi} \right) = 0 \quad (4)$$

where $c_p = \sqrt{\frac{E}{\rho}}$ - speed of propagation of longitudinal waves in the ξ-direction,

$f_s(\xi) = f_r^2(\xi)$ - function describing the change of waveguide cross section in the direction ξ,
$f_r(\xi)$ - function describing the change of cross section radius in the direction ξ.

The solution (4) can be assumed in the form

$$\overline{u}(\xi,t) = \overline{U}(\xi)T(t), \quad (5)$$

where function $\overline{U}(\xi)$ represents the mode shape of the longitudinal vibration of the waveguide and the function $T(t)$ describes the time dependence of the waveguide longitudinal vibration.

The structural design of ultrasonic cone waveguide with adaptive modal properties is presented in Figure 3. The cone waveguide has input radius R_i, output radius R_o and length L_o. The ultrasonic cone waveguide body has a drilled hole (radius R_c) for insertion of core with the length L_c. Position of core in waveguide body from input radius is defined by dimension L_s. The different material properties will be used for body of cone waveguide and embedded core. The longitudinal displacements of interacting points between the cone waveguide body and the core are the same, i.e. perfect adhesion is assumed for the corresponding points in interface between core and waveguide body.

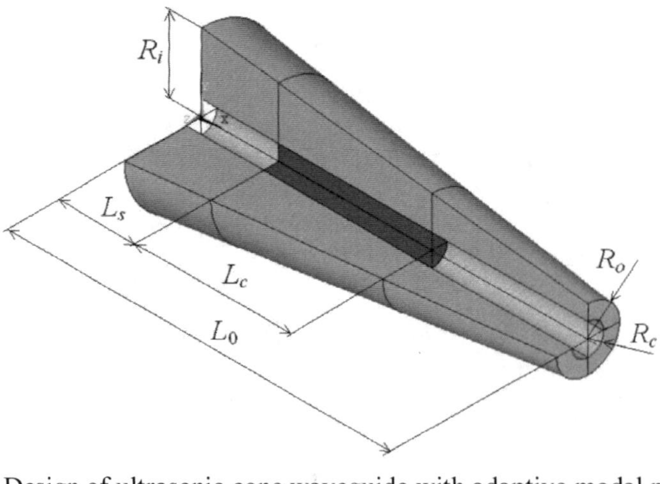

Figure 3. Design of ultrasonic cone waveguide with adaptive modal properties.

IOP Conf. Series: Materials Science and Engineering 1050 (2021) 012009 doi:10.1088/1757-899X/1050/1/012009

In order to generalize the results of the analyzes, the following geometric dimensionless parameters and material properties are introduced

- dimensionless input radius of waveguide cross section

$$\delta_R = \frac{R_i}{L_0} \quad (\approx R_i) \tag{6}$$

- dimensionless output radius of waveguide cross section

$$\delta_o \delta_R = \frac{R_o}{L_0} \quad (\approx R_o), \qquad \text{where} \qquad \delta_o = \frac{R_o}{R_i} \tag{7}$$

- dimensionless radius of core cross section

$$\delta_c \delta_R = \frac{R_c}{L_0} \quad (\approx R_c), \qquad \text{where} \qquad \delta_c = \frac{R_c}{R_i} \tag{8}$$

- dimensionless core position

$$\xi_s = \frac{L_s}{L_0} \quad (\approx L_s), \tag{9}$$

- dimensionless core length

$$\xi_c = \frac{L_c}{L_0} \quad (\approx L_c), \tag{10}$$

- dimensionless density of core material

$$\kappa_\rho = \frac{\rho_c}{\rho}, \tag{11}$$

- dimensionless Young modulus of core material

$$\kappa_E = \frac{E_c}{E}. \tag{12}$$

The dimensionless geometric parameters for conical shape of the waveguide with the inserted reinforcing core are shown in Figure 4.

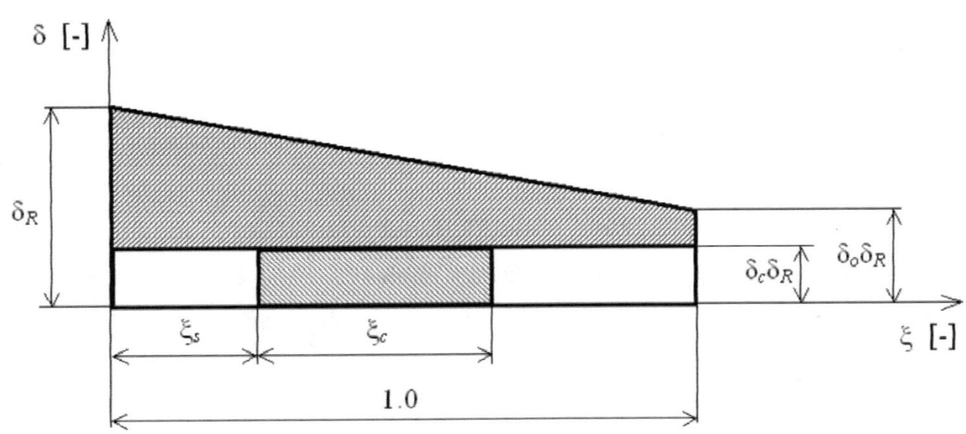

Figure 4. Dimensionless geometrical parameters of cone waveguide with inserted core.

The finite element method (ANSYS software code environment) is used to create a computational model and perform a modal analysis of a conical waveguide and determine its modal properties (mode shapes, natural frequency, amplification factor). Typical mode shapes (half wave and full wave) of waveguide vibrations in resonant states are shown in Figure 5. The input and output amplitudes of the mode shape are expressed as follows

$$A_i = \overline{U}(\xi = 0) \qquad \text{and} \qquad A_o = \overline{U}(\xi = 1)\,. \tag{13}$$

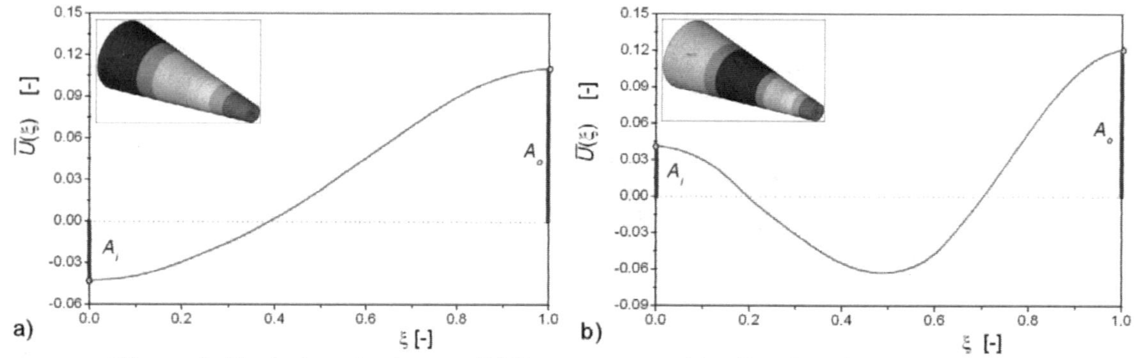

Figure 5. Typical mode shapes of full cone waveguide vibrations in resonant state.
a) half wave mode shape; b) full wave mode shape

The cone waveguide with adaptive modal properties (Figure 3, Figure 4) has similar mode shapes to those shown for full cone waveguide (Figure 5) without the possibility of modifying modal properties. The performance and efficiency of a waveguide with adaptive modification of modal properties are evaluated using the values of its modal properties, which are modified due to the change of geometric parameters and material properties of the inserted core.

In order to compare the effect of structural modifications of cone waveguide on its modal properties and to generalize the results of the analyzes, the dimensionless modal properties such as dimensionless natural frequency and dimensionless amplification factor are formulated, as follows

• the first dimensionless natural frequency

$$\psi_1 = \frac{f_{01,m}}{f_{01}}, \tag{13}$$

• the second dimensionless natural frequency

$$\psi_2 = \frac{f_{02,m}}{f_{02}}, \tag{14}$$

• the first dimensionless amplification factor

$$\theta_1 = \frac{A_{1,o}}{A_{1,i}}, \tag{15}$$

• the second dimensionless amplification factor

$$\theta_2 = \frac{A_{2,o}}{A_{2,i}}, \tag{16}$$

where

$f_{01,m}$, ($f_{02,m}$) - the first (second) natural frequency of modified waveguide structure (with core),
f_{01}, (f_{02}) - the first natural frequency of original waveguide structure (without core),
$A_{1,i}$, ($A_{2,i}$) - input amplitude for the first (second) mode shape of waveguide,
$A_{1,o}$, ($A_{2,o}$) - output amplitude for the first (second) mode shape of waveguide.

3. Numerical analysis and results

The computational models and half-wave and full-wave mode shapes of cone waveguide vibrations with the possibility of adaptive modification of modal properties, by inserting the core into the waveguide body, for different core positions, are shown in Figure 6.

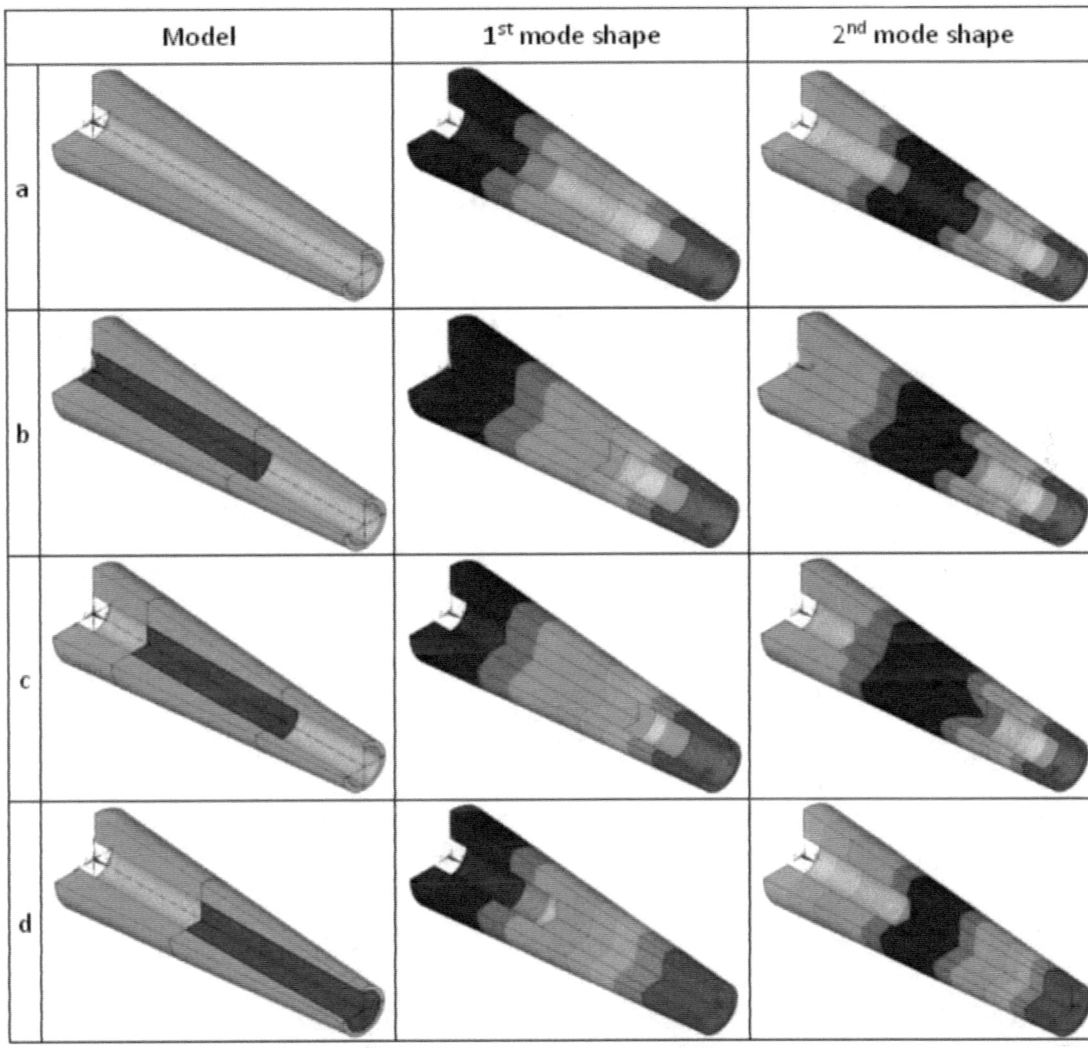

Figure 6. Mode shapes of cone waveguide vibrations with adaptive modification modal properties.

The effect of structural changes of the cone waveguide by the inserted core on the first and second natural frequencies are graphically presented in Figures 7-8. Similarly, the effect of the presented structural changes cone waveguide by the inserted nucleus on the first and second amplification factors are presented in Figures 9-10. In these graphs, the effects of changes in the position and length of the core in the cone waveguide body for different dimensionless densities and young modulus of elasticity are presented.

The dependencies of dimensionless natural frequencies and dimensionless amplification factors on the change in structural parameters for the case when the reinforcing core is gradually inserted into the waveguide body (Figure 6) until the entire hole in the waveguide body is filled with the core are shown in Figure 7 and Figure 8. For the structural parameters that have been considered and that can be changed, it is possible to modify the first dimensionless natural frequency in the range from –16 %

to +16 % and the second dimensionless natural frequency in the range from −12 % to +16 %. For the first dimensionless amplification factor, the range of the modification is from –26 % to +44 %, and for the second dimensionless amplification factor, the range is from –28 % to 75 %.

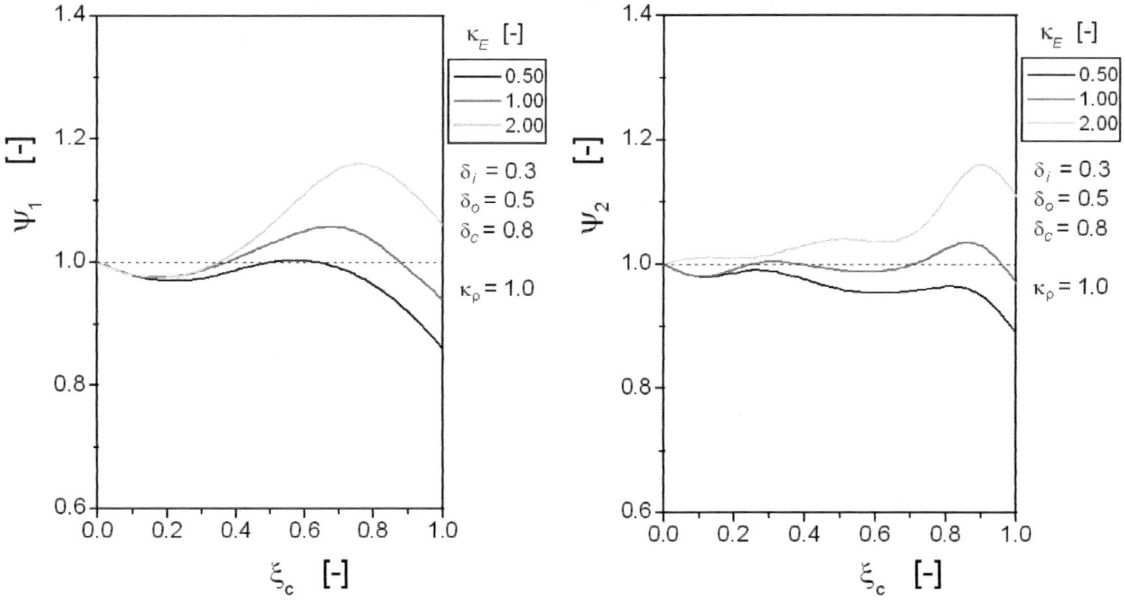

Figure 7. Dependency of the first (a) and second (b) natural frequencies on the dimensionless core length for parameters $\kappa_E = \{0.5; 1.0; 2.0\}$.

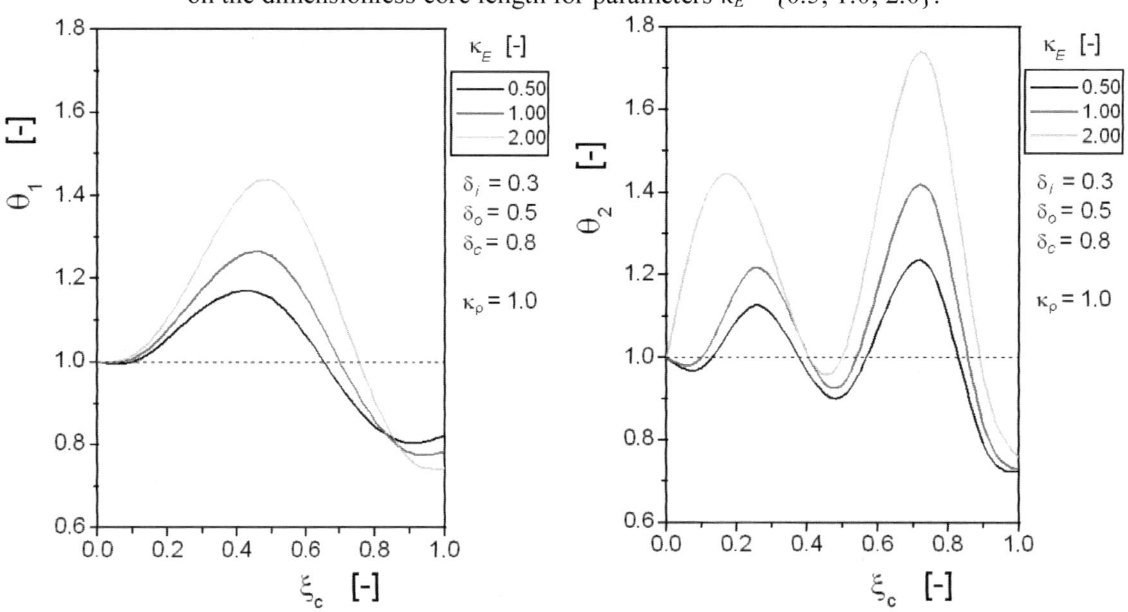

Figure 8. Dependency of the first (a) and second (b) amplification factors on the dimensionless core length for parameters $\kappa_E = \{0.5; 1.0; 2.0\}$.

The second case for the modification of the modal properties of a conical waveguide, which concerns the insertion of a reinforcing core of finite length into the waveguide body (Figure 6c), where two dimensionless core lengths ξ_c (0.3 and 0.5) were considered. The modifications of dimensionless natural frequencies and dimensionless amplification factors are shown in Figure 9 and Figure 10.

When the considered structural parameters are modified, the modification of the first dimensionless natural frequency in the range from −13 % to +7 % (for ξ_c = 0.3), respectively from −13 % to +4 % (for ξ_c = 0.5) and the second dimensionless natural frequency in the range from −7 % to +3 % (for ξ_c = 0.3), respectively from −1.0 % to +1.2 % (for ξ_c = 0.5). For the first dimensionless amplification factor, the range of the modification is from −40 % to +1 % (for ξ_c = 0.3), respectively from −51 % to 0 % (for ξ_c = 0.5), and for the second dimensionless amplification factor, the range of the modification is from −59 % to +17 % (for ξ_c = 0.3), respectively from −20 % to +16 % (for ξ_c = 0.5).

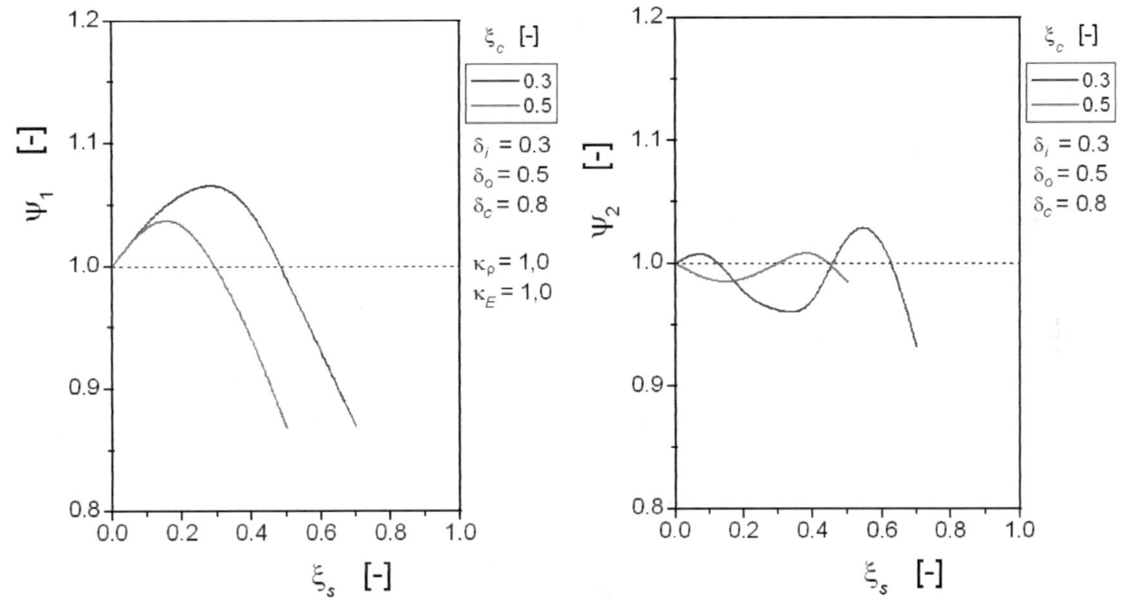

Figure 9. Dependency of the first (a) and second (b) natural frequencies on the dimensionless core position ξ_s and specific dimensionless core length ξ_c = {0.3; 0.5}.

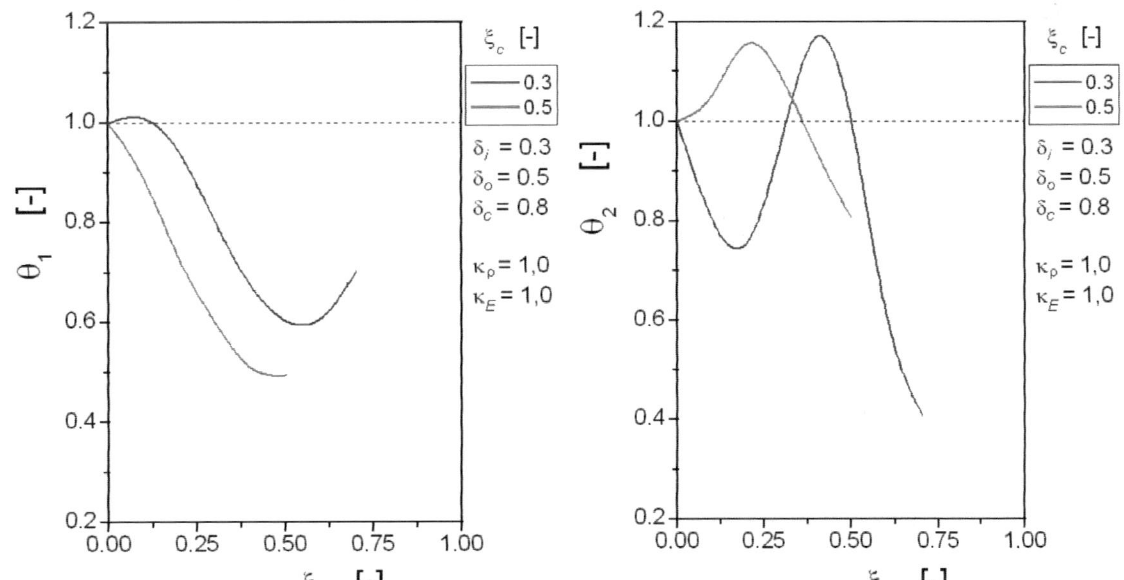

Figure 10. Dependency of the first (a) and second (b) amplification factors on the dimensionless core position ξ_s and specific dimensionless core length ξ_c = {0.3; 0.5}.

Conclusions

Method for modification (tunning) modal properties (natural frequencies and amplification factor) of cone waveguide structure for excitation system for ultrasonic assisted technological processes is presented. The method of modal properties modification of waveguide is based on structural modification of waveguide parameters (i.e. redistribution of mass and stiffness properties), which consists in inserting a reinforcing core into the waveguide body.

By inserting a reinforcing core, it is possible to achieve relatively significant modification of modal properties (natural frequencies and amplification factors) depending on the length of inserted core, on the position of inserted core in the waveguide body and the core material properties.

Based on the results obtained, it can be concluded that this method of modifying the modal properties provides suitable possibilities for "tuning" the conical waveguide for the needs of performing the technological process in the optimal mode.

Acknowledgments

The research work was supported by the project KEGA 029STU-4/201 and the projects of the Slovak University of Technology in Bratislava for “Young researchers” - MTF 1355 and MTF 1357.

References

[1] Astashev V K and Babitsky V I 2007 *Ultrasonic Processes and Machines: Dynamics, Control and Applications* (New York: Springer Science+Business Media).

[2] Babitsky V I, Kalashnikov A N and Molodtsov F V 2004 Autoresonant control of ultrasonically assisted cutting *Mechatronics* **14** (1) 91-114.

[3] Brehl D E and Dow T A 2008 Review of vibration-assisted machining *Precis. Eng.* **32** (3) 153-172.

[4] El-Hofy H A G 2005 *Advanced Machining Processes: Nontraditional and Hybrid Machining Processes* (New York, The McGraw-Hill Companies).

[5] Dekys V Saga M and Zmindak M 2004 *Dynamics and Reliability of Mechanical Systems* (VTS ZU, Zilina). (in Slovak)

[6] Thorby D 2008 *Structural Dynamics and Vibration in Practice - An Engineering Handbook* (Elsevier Ltd., Oxford)

Comparison of microstructures and mechanical properties of Al-Ti weld joints prepared by different welding technologies

M Nagy, M Behúlová and Š Vrtiel

Slovak University of Technology in Bratislava, Faculty of Materials Science and Technology in Trnava, Ulica Jána Bottu 2781/25, 917 24 Trnava, Slovak Republic

mate.nagy@stuba.sk, maria.behulova@stuba.sk, stefan.vrtiel@stuba.sk

Abstract. This study deals with the evaluation of butt welded-brazed joints of two plates with the thickness of 2 mm from EN AW5083-H111 alloy and titanium Grade 2 prepared by two different welding technologies: laser beam welding-brazing (LBWB) and gas tungsten arc welding (GTAW). ER4047 and ER4043 welding wires with the diameter of 1.2 mm were used as filler materials. The macrostructures, microstructures and mechanical properties of experimental weld joints were analysed and compared. In general, the strength of joints produced by LBWB was higher than the strength of GTAW samples. The highest tensile strength at the level of 245 MPa was measured for the experimental joint made by LBWB using following welding parameters: the laser power of 1800 W, the welding speed of 30 $mm.s^{-1}$, the laser beam offset to Al-sheet of 400 μm and the wire feed rate of 2.4 $m.min^{-1}$.

1. Introduction

The difficulties in joining aluminium to titanium alloys by fusion welding technologies result not only from significant differences in the physical, mechanical and thermal properties of the two materials but mainly from the unavoidable formation of layers containing brittle intermetallic compounds [1-4]. Various conventional and unconventional welding technologies were tested to obtain sound weld joints of these materials, and the influence of welding technologies on the quality of produced weld joints was evaluated [5-6]. Because of GTAW welding technology offers flexibility, low dependence on expensive equipment with complex welding procedures it is widely used in manufacturing application. GTAW welding is characterized by relatively small temperature gradients, which contributes to improving the homogeneity of the interfacial reaction. Published research studies related to the GTAW welding-brazing of dissimilar Al/Ti materials have reported that the application of the AlSi12 filler metal with the high Si content supressed the growth of intermetallic compounds (IMC), thereby enhancing the mechanical properties of produced joints [7-10]. Many other studies showed the promising results related to Al/Ti laser beam welding-brazing (LBWB) [11-14]. When using this joining method, the molten aluminum alloy only wets the solid surface of the titanium. On the other hand, further investigations have demonstrated that the insufficient width, discontinuity or improper morphology of an interfacial reaction layer could cause reduction of mechanical properties and initiation of joint cracks. Similarly, it was proved that the laser offset influenced the thickness of the IMC layer and its application could improve the mechanical properties of the weld by eliminating the interfacial non-homogeneity and weld porosity [14-15].

The aim of this paper is to evaluate the differences in macrostructure, microstructure and mechanical properties of butt welded-brazed joints of Al/Ti dissimilar materials produced by two different welding technologies using AlSi based filler metals.

2. Experimental materials, equipment and welding technologies

Butt welded-brazed joints of dissimilar experimental base materials, wrought EN AW5083-H111 (AA5083) aluminium alloy and titanium Grade 2 (CP Grade 2) with chemical compositions given in Table 1 were prepared using different welding technologies - laser beam welding-brazing and gas tungsten arc welding. Experimental samples had dimensions of 50 mm × 50 mm × 2 mm. A filler material in the form of a rod and wire, respectively, with the diameter of 1.2 mm was applied. In both technologies, argon was used as a shielding gas with the flow rate of 20 $l.min^{-1}$. The laser beam and the welding arc, respectively, was deflected to the Al sheet side to reduce the temperature rise of the CP Grade 2 and its possible melting. Experimental robotized welding device, the TruDisk 4002 disk laser was used in continuous mode. The laser beam with the diameter of 400 μm was focused +2 mm above the horizontal reference plane. The laser welding-brazing was accomplished using ER4047 (AlSi12) filler wire and the following welding parameters: laser power of 1800 W, welding speed of 30 $mm.s^{-1}$, wire feed rate of 240 $cm.min^{-1}$ and the laser beam offset of 400 μm to Al sheet side with respect to the weld centreline.

For selected conventional welding technology GTAW, the SELCO Genesis 3200 AD/DC inverter-type welding machine was exploited. The 2-step welding method with direct current was applied to produce a double-square-groove type of joint in flat position using ER4043 (AlSi5) filler material with the chemical composition listed in Table 1. The welding speed was 2.5 $mm.s^{-1}$ along with the arc voltage of (12.8 ± 0.4) V and the welding current of 50 A. The arc offset was approximately 400 μm to Al sheet side.

Table 1. Chemical composition of experimental materials in [at. %].

Element	Mg	Si	Fe	Mn	Cu	Zn	Al	C	O	N	H	Ti
CP Grade2	-	-	0.25	-	-	-	-	0.393	0.737	0.101	0.702	Bal.
AA5083	5.22	0.38	0.15	0.13	0.08	-	Bal.	-	-	-	-	-
AlSi12	0.11	11.67	0.39	0.08	0.13	0.08	Bal.	-	-	-	-	-
AlSi5	0.06	5.09	0.39	0.03	0.13	0.04	Bal.	-	-	-	-	-

Measured mechanical properties of experimental base materials and mechanical properties of the filler materials according to material datasheets are summarized in Table 2 [16-17]. The AA5083 base Al alloy with the main alloying element Mg was strain hardened and cold rolled. The analysed microstructure of the experimental alloy AA5083 was homogeneous, with the grains slightly elongated in the rolling direction. The grains reached an average size of (35 ± 4) μm in the rolling direction and (27 ± 2) μm in the direction transverse to the rolling direction. The microstructure of CP Grade 2 consists of fine equiaxed α grains with an average grain size diameter of (22 ± 2) μm. The experimental material was cold strain hardened [16].

Table 2. Mechanical properties of experimental materials.

Material	Tensile strength [MPa]	Young modulus [MPa]	Elongation [%]	Microhardness [HV0.1]	Ref.
CP Grade2	330 ± 13	105	20	170 ± 5	[16]
AA5083	267 ± 15	73.4	29.7 ± 1	72 ± 2	[16]
AlSi12	290 - 320	79	15 - 22	120 - 140	[17]
AlSi5	145 - 227	79	5 - 12	-	[17]

Test specimens prepared from experimentally produced welded-brazed joints were examined using mechanical testing, macrostructural and microstructural observations including chemical EDX analysis and fractographic evaluation. The destructive transverse tensile tests were performed on LabTest 5.250 SP1-VM testing machine with testing speed of 1 $mm.s^{-1}$, according to standards STN EN ISO 6829-1 [18] and STN EN ISO 4136 [19]. The microhardness was measured by Buehler IndentaMet 1100 microhardness testing machine using testing load HV 0.1 during 10 s in accordance with STN EN ISO 6507-1 [20] and STN EN ISO 9015-2 [21]. Specimens for macroscopic and microscopic examinations were prepared from cross-sections perpendicular to the weld centreline and etched with Keller´s etchant (consisting in volume and in order of mixing from 950 ml water, 15 ml hydrochloric, 10 ml hydrofluoric acid) for 3-5 seconds [22]. ZEISS Stemi 2000-C equipment with AxioCam ERc5s camera and NEOPHOT 32 light microscope were used for macroscopic and microscopic examination of specimens. Chemical compositions at the joint interfaces were measured using JEOL 7600 scanning electron microscope equipped with EDX analyser.

3. Experimental results

Comparison of welded-brazed joints produced by conventional and unconventional welding technologies was accomplished on the base of measured mechanical properties, differences in welded-brazed joints macrostructure and microstructure, chemical composition of WM-Ti interface and the examination of fracture surfaces after transverse tensile strength tests. Using similar filler metal AlSi12 with LBWB and AlSi5 at double-square-groove type of joint produced with GTAW, the influence of the Si content was neglected. The heat input in both welding technologies was deflected to the Al sheet to reduce the amount of Ti melting at the joint interface.

3.1. Macrostructural and microstructural analysis of welded-brazed joints

The macrostructures of welded-brazed joints prepared by laser beam and GTAW welding-brazing technologies are compared in Figure 1. Significant differences can be clearly seen in the dimensions of WM. The cross-sectional area of WM is depended mainly on the heat input to the joint. Using GTAW welding technology, the linear heat input of 190.5 $J.mm^{-1}$ was supplied into the joint, thus obtaining a WM area with a relatively large size of 21.34 mm^2 (Figure 1b). It is also necessary to notice that by GTAW welding technology a double-square-groove type of joint was produced. In contrast to conventional GTAW welding technology, in unconventional LBWB welding technology, the linear heat input of approximately 45 $J.mm^{-1}$ was applied to the joint, producing a WM cross-sectional area of 6.20 mm^2 (Figure 1a).

Detailed evaluation of experimental joints macrostructures proved that the areas of joint interface fit for typically welded-brazed joints cross-sections, where two different zones are formed. From the Ti side, where the joint is partially melted, a Ti-WM brazed interface was developed. In the WM zone, complete melting and mixing of the AA5083 base material and the filler metal took place.

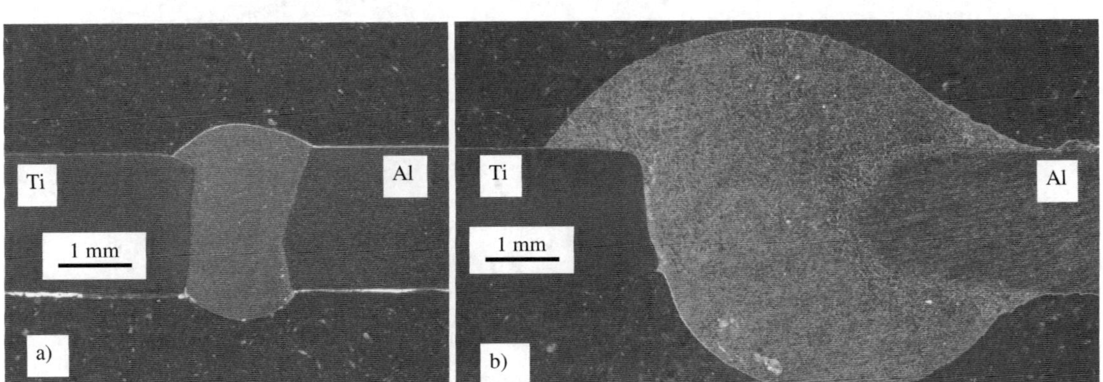

Figure 1. Macrostructures of experimental joints produced by a) LBWB and b) GTAW.

Figure 2 shows the microstructure of welded-brazed joints after LBWB and GTAW at the Ti-WM interface. The WM microstructure consists of quasi-globular grains (Figures 2a and 2c) formed by mixing the filler material (AlSi12 or AlSi5) and AA5083 Al alloy and following solidification of the melt. It is assumed that the WM consists of a solid solution of α, where the interdendritic spaces are enriched with alloying elements Si and Mg creating fine precipitates due to segregation. The finer-grained globular shaped microstructure in Figure 2a was developed due to the rapid temperature cycle (rapid heating and rapid cooling of materials) characteristic for the laser welding technology.

A more detailed comparison of the Ti-WM interface of welded-brazed joints experimentally produced by different technologies is shown in Figures 2b and 2d. Generally, if the heat supplied to the weld during the Al-Ti joining process is sufficient to partially melt the titanium sheet, the diffusion of Ti, Al and other atoms of the alloying elements occurs at the Ti-WM interface, resulting in the formation of different layers of intermetallic compounds. From this point of view, a typical welded-brazed joint was developed at the Ti-WM interface during GTAW welding, where the melting of CP Grade 2 was minimal and not continuous along the entire Ti front surface. As a result of these conditions, one thin discontinuous IMC layer was formed at the Ti-WM interface with an average thickness of 1 μm (Figure 2d). On the other hand, the Ti-WM interface of the LBWB joint consists of one continuous and one quasi-continuous IMC layer with a total thickness of 10.2 μm (Figure 2b).

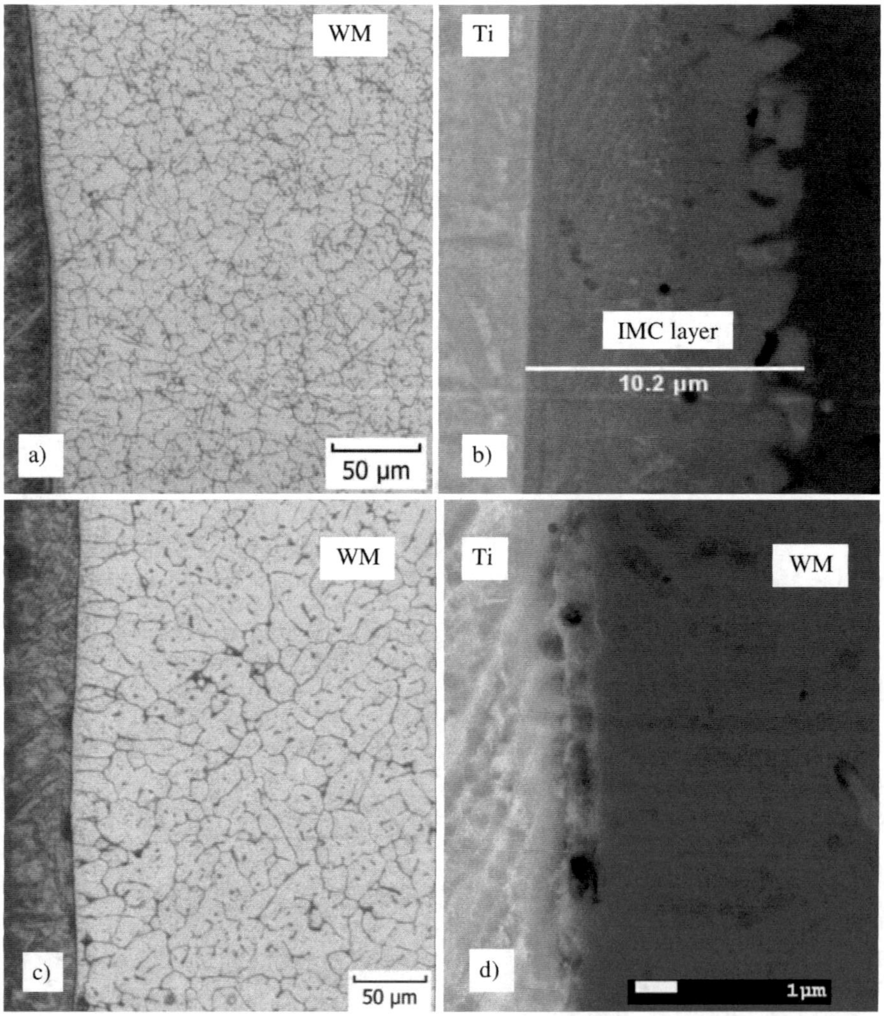

Figure 2. Microstructures of joints at Ti-WM interface prepared by a,b) LBWB and c,d) GTAW.

3.2. EDX chemical analysis

Figure 3 shows a detail from the Ti-WM interface of an experimental welded-brazed joint prepared by LBWB technology with the points selected for the EDX analysis. At the Ti-WM interface of this joint two kinds of IMC layers have been formed, one continuous layer followed by another quasi-continuous layer (in the direction from the Ti side to WM).

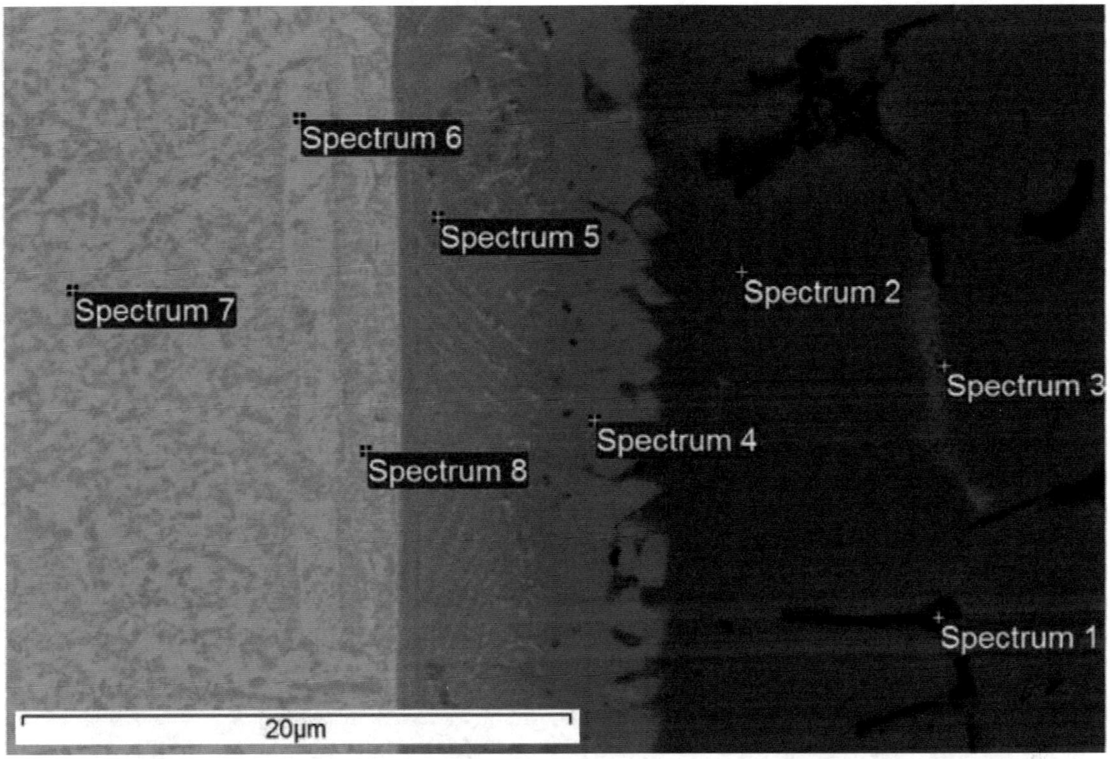

Figure 3. Positions of EDX analysis at the Ti-WM interface of a joint produced by LBWB.

Results of EDX analysis for the spectra 1-8 are summarized in Table 3. Based on the chemical composition, phase diagrams and the literature sources [23-25], possible phases present were predicted. Spectra 1-3 are characterized by a high Al content typical for WM. In the darker interdendritic spaces in spectrum 1, a higher atomic percentage of Mg in the Al matrix was identified in comparison with spectrum 3, where the amount of Si [at. %] increases due to segregation and the Mg content decreases. The chemical composition of the discontinuous IMC layer at the Ti-WM interface is represented by the spectrum 4. The measured Ti content and Al content is 24 at. % and 72 at. %, respectively, what corresponds to the composition of $TiAl_3$ intermetallic compound.

Table 3. EDX analysis [at. %] in the points according to Figure 3.

Spectrum	Al	Ti	Si	Mg	Mn	Fe	Possible phases [23-25]
1	90.87	0.14	7.37	1.51	0.06	0.06	Al matrix
2	97.20	0.43	0.77	1.55	0.06	0.00	Al matrix
3	89.02	0.11	8.29	0.76	0.22	0.96	Al matrix
4	71.44	23.94	4.56	0.00	0.00	0.22	$TiAl_3$
5	47.76	43.06	9.18	0.00	0.00	0.07	TiAl
6	0.43	99.42	0.15	0.00	0.00	0.00	αTi
7	0.24	99.76	0.00	0.00	0.00	0.00	αTi
8	2.35	97.48	0.16	0.00	0.00	0.00	αTi

The chemical composition of the continuous IMC layer located closer to Ti is represented by spectrum 5 with the Al content of 47.76 at. % and the Ti content of 43.03 at. %. That could indicate the presence of AlTi IMC. The chemical compositions of spectra 6-8 are very similar to the chemical composition of the base material CP Grade 2. In these spectra, a slight decrease of Al concentration with an increasing distance of the point from the Ti-WM interface towards to CP Grade 2 can be observed.

A detail from the area of Ti-WM interface of a welded-brazed joint produced by GTAW technology is shown in Figure 4 with the spectra where the chemical composition was measured using EDX analysis. In contrast to the weld prepared by LBWB technology, in the case of GTAW welding, only one thin discontinuous layer of intermetallic compound was formed at the Ti-WM interface. Table 4 summarizes the chemical composition in selected points according to Figure 4, i. e. in WM (spectra 1 - 4 and 8), at the interface of Ti-WM (spectra 5 and 7) and in the HAZ of the CP Grade 2 (spectrum 6). The possible phases present were predicted in this case as well. Spectra 2-4 and 8 exhibited high content of Si element (above 12 at. %) and suggested development of Al-Si eutectics with a silumin-like structure.

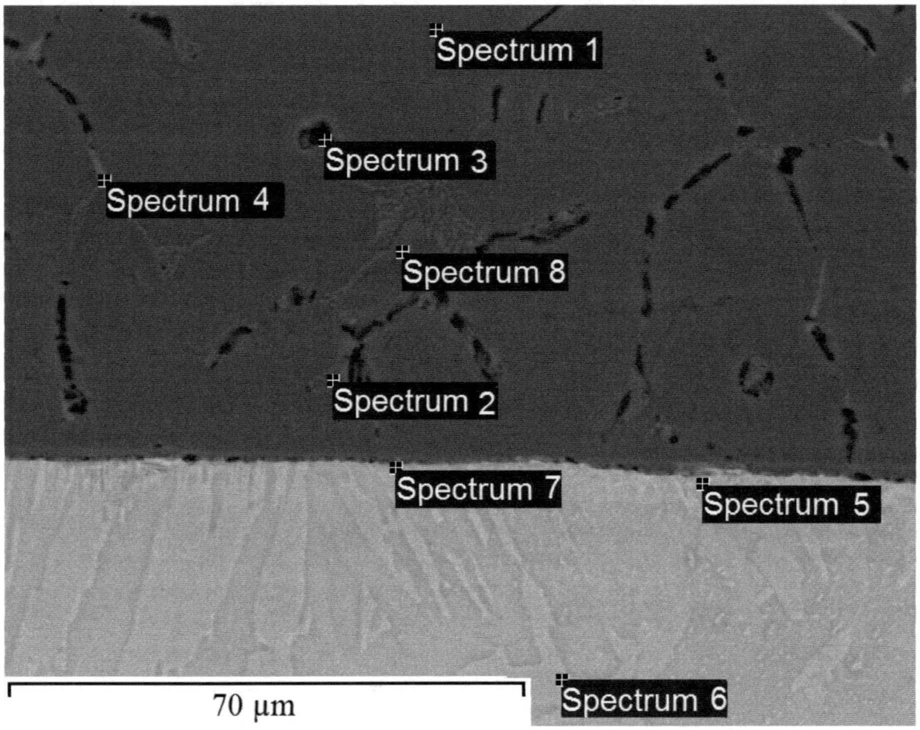

Figure 4. Positions of EDX spectra at the Ti-WM interface of a joint produced GTAW.

Table 4. EDX analysis [at. %] in the points according to Figure 4.

Spectrum	Al	Ti	Si	Mg	Fe	Possible phases [23-25]
1	97.8	0.11	0.97	0.21	0.00	Al matrix
2	63.19	0.08	32.76	3.27	0.20	Al-Si
3	86.30	0.00	12.63	0.56	0.00	Al-Si
4	69.49	0.00	19.71	8.10	2.18	Mg_2Al_3, Al-Si
5	6.14	92.83	0.77	0.00	0.26	Ti_3Al, Ti_3Si
6	0.47	99.53	0.00	0.00	0.00	αTi
7	42.08	55.87	1.88	0.00	0.00	Ti-Al
8	76.69	0.10	22.30	0.34	0.00	Al-Si

3.3. Mechanical properties of welded-brazed joints

The course of micro-hardness was measured on the cross-sections along the centre of the sheet thickness with spacing of 250 μm between the indentations. In Figure 5a, the measured values of microhardness for welded-brazed joints produced by different welding technologies are compared. The width of the weld metal for the joint prepared by LBWB technology (≈ 2.4 mm) is approximately two times lower in comparison with WM width of the joint produced by GTAW method (4.9 mm). On the other hand, the maximum values of WM microhardness are slightly higher for LBWB joint (max. 110 HV0.1) than the comparative values of microhardness in WM of joint produced by GTAW technology (max. 90 HV0.1). This trend can also be recognized in heat-affected zones (HAZ), where the difference in microhardness is approximately 15%, while the difference in HAZ width is even greater (HAZ LBWB width ≈ 2.1 mm and HAZ GTAW width > 5 mm).

The increase in the microhardness values of the weld metal by almost 40 % compared to the microhardness values of the base material (AA5083) for joints prepared by LBWB may be due to grain refinement resulting from the rapid temperature cycle and mixing the base material with the filler metal. On the Ti side, at the interface and in the HAZ, there is an increase of the microhardness compared to the microhardness of the CP Grade 2 base material due to the polymorphic transformation of Ti. The width of the HAZ on the Ti side is approximately two times smaller than the HAZ width on the Al side.

The microhardness decrease in the HAZ of the GTAW joint could be caused by grain coarsening and diffusion of Mg from the Al matrix. Fine, recrystallized Al grains, enriched mainly with silicon, forms eutectics at the grain boundaries of Al-Si, which increase the microhardness values of the weld metal by consolidating the solid solution α. The microhardness values HV0.1 at the WM rise above the measured microhardness values of the base Al alloy. At Ti side, there was a decrease of microhardness values in contrast to the Ti base material values.

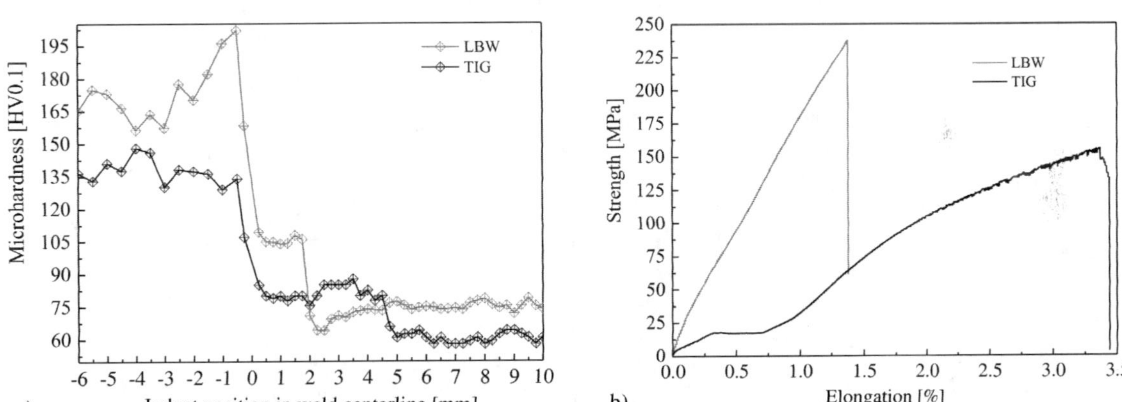

Figure 5. Results of mechanical testing of experimental joints a) microhardness measurements across the weld centreline and b) transversal tensile strength tests.

Figure 5b illustrates the results of transversal tensile tests of welded-brazed joints prepared by LBWB and GTAW technologies. The ultimate tensile strength of test specimens from joints produced by LBWB reached the value of 245 MPa, what is nearly 90 % of the tensile strength of the AA5083 base material. Experimental specimens prepared from GTAW joints reached only 57 % of the ultimate tensile strength of the Al base material, which was 153 MPa. At both technologies, the fracture occurred in the Ti-WM interface. The welded-brazed joints prepared by GTAW technology exhibited better elongation values at failure than the joint prepared by LBWB. This may be due to the type of joint prepared by GTAW technology (butt double-square-groove type of joint), in which a larger volume of a filler metal is used.

IOP Conf. Series: Materials Science and Engineering 1050 (2021) 012010 doi:10.1088/1757-899X/1050/1/012010

3.4. Fractographic analysis of welded-brazed joints

The mode of fracture of Ti–Al welds depends on the morphology and phase composition of the interface [26]. The fracture of transversal tension test specimen prepared by LBWB technology took place in the Ti-WM interface. Both fracture surfaces (at WM side and Ti side) documented in Figure 6 were evaluated. Figures 6a and 6b illustrate the fracture surface at the WM side where the cleavage cracks and transcrystalline fracture trace can be observed. From the energetic point of view, this fracture can be considered as a brittle fracture with low fracture energy, which is characteristic for cleavage cracks. The fracture surface at the Ti side in Figure 6c with a detail in Figure 6d is very similar to the fracture surface at WM. The entire investigated fracture surface was melted on the Ti-side. During the melt solidification, microcavities developed which acted as stress concentrators. Cleavage cracks were initiated mainly in the zones containing microcavities and/or discontinuous IMC layers. In the right upper corner in Figure 6c, cleavage facets in groups, typical for a brittle fracture can be seen.

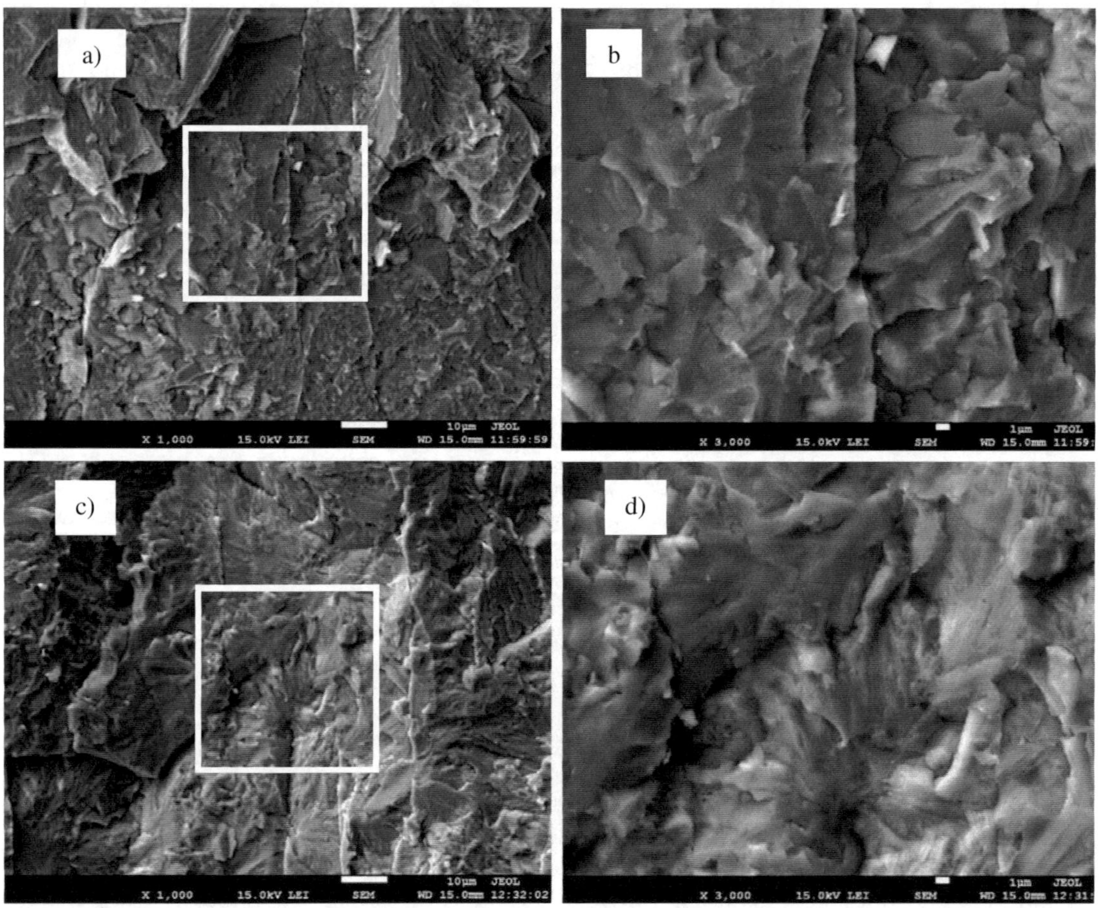

Figure 6. SEM photographs of fracture surfaces of transverse tensile test specimen prepared from LBWB joint a, b) at WM side and c, d) at Ti side.

The fracture of transverse tension test specimen prepared from welded-brazed joint produced by GTAW technology occurred as well as at the WM-Ti interface. Figure 7a shows the fracture surface at the weld metal side, which is formed by the Al matrix and Al-Si eutectics. It is possible to observe cavities and pores developed due to the trapped gases (especially hydrogen), which could not escape during solidification. The pores and impurities in the WM were most likely the initiators of defects, the growth of which caused a weld failure. A detail in Figure 7b confirms a mixed type of the obtained fracture. Based on the trace of the fracture in the lower left corner, the fracture can be classified as a predominantly intercrystalline quasi-cleavage fracture, which changes into a transcrystalline ductile fracture in the upper right corner.

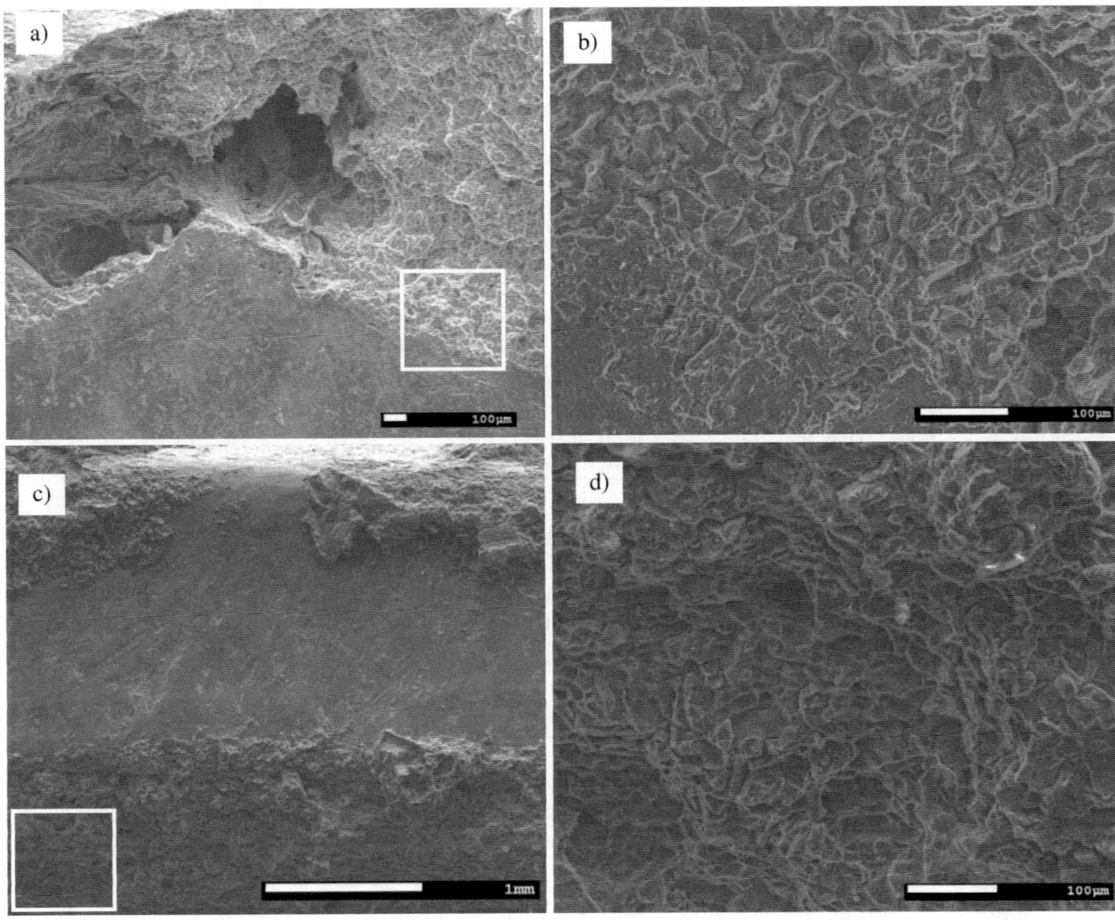

Figure 7. SEM photographs of fracture surfaces of transverse tensile test specimen prepared from GTAW joint a, b) at WM side and c, d) at Ti side.

The fracture surface at the Ti side, where the joint was formed mainly with brazing, exhibits mixed character (Figures 7c and 7d). The surface of unmelted CP Grade 2 sheet was only wetted with the molten WM consisting of a mixture of the AA5083 alloy and the filler metal (Figure 7c). Despite the higher heat input during the GTAW welding and the double-square-groove type of joint, the Ti side was only partially melted. This fact was probably caused by a larger deflection of the welding arc to the Al side during manual GTAW welding process compared to the automatized LBWB process. The surface and the root of the fracture surface on the Ti side shows more-less ductile fracture in contrast to the brittle fracture of the LBWB specimen. In a detail in Figure 7d, a brittle, low-energy ductile failure can be seen, which also indicates the extent of plastic deformation of the fracture surface. From the morphological point of view, the fracture surface on the Ti side corresponds to a quasi-cleavage fracture in which microcracks are located.

4. Conclusions

The paper presents the comparison of macrostructures, microstructures and mechanical properties of welded-brazed Al/Ti joints prepared by different welding technologies. Butt joints of dissimilar experimental base materials, wrought AA5083 aluminium alloy and titanium Grade 2 were produced by unconventional laser beam welding-brazing technology and conventional gas tungsten arc welding technology. Plates with the thickness of 2 mm were butt joined using AlSi12 and AlSi5 filler materials.

The welded-brazed joint prepared by LBWB technology exhibits higher strength at the level of 245 MPa compared to the strength of GTAW sample. The width of WM was two times lower for samples produced by LBWB technology, than in the GTAW samples. This trend was confirmed also by the microhardness measurements on the joint cross-sections along their horizontal centreline.

The macroscopic analysis of joints prepared by both welding technologies shows the typical welded-brazed characteristics of joint at the Ti-WM interface. When comparing the microstructures of WM near the Ti interface of experimentally produced joints, the differences can be found in the grain size, grain morphology and the degree of mixing the experimental materials. In both cases, the weld metal consists of α solid solution and the Al-Si eutectics located in the interdendritic spaces but the finer-grained microstructure was developed in the WM of joint produced by LBWB technology. In addition, IMF layers with the total thickness of 10.2 μm were formed at the Ti-WM interface due to the melting of Ti surface and occurrence of diffusion processes at the Ti-WM interface. Chemical EDX analysis of the Ti-WM interface of LBWB joint using a higher resolution revealed one continuous IMC layer with the chemical composition corresponding to the TiAl IMC followed by a discontinuous IMC layer with the composition similar to $TiAl_3$. In case of GTAW joint, only one thin discontinuous IMC layer was formed at the Ti-WM interface with an average thickness of 1 μm.

The fractographic analysis showed that the fracture of transverse tension test specimen prepared from the joint produced by LBWB technology occurred by a brittle fracture with a transcrystalline fracture traces. The fracture surfaces of the GTAW joint exhibit a mixed character of the fracture resulting probably from insufficient melting of the Ti surface.

Acknowledgements
The research has been supported by the Scientific Grant Agency of the Slovak Republic within the Project VEGA No. 1/0796/20.

References

[1] Kalaiselvan K and Elango A 2014 *JAMME* **67** 1 39-44.
[2] Sánchez-Amaya J M, Amaya-Vázquez M R and Botana F J *2013* Laser welding of light metal alloys: aluminium and titanium alloys S. Katayama (Ed.) *Handbook of Laser Welding Technologies* (Woodhead Publishing Series in Electronic and Optical Materials) https://doi.org/10.1533/9780857098771.2.215.
[3] Ohnuma I et al 2000 *Acta Mater.* **48** 3113-23 https://doi.org/10.1016/S1359-6454(00)00118-X
[4] Zhang Y et al 2018 *J. Alloys Compd.* **747** 764-71 https://doi.org/10.1016/j.jallcom.2018.03.119.
[5] Xu C and Peng C 2020 *Mater. Res. Express* **7** 026542.
[6] Zhou X, Duan J, Zhang F and Zhong S 2019 *Materials* **12** (5) 712 https://doi.org/10.3390/ma12050712.
[7] Liu K, Li Y, Wei S and Wang J 2014 *Mater. Manuf. Proces.* **29** 969-974 https://doi.org/10.1080/10426914.2013.864414.
[8] Wang H, Yuan X, Li T, Wu K, Sun Y and Xu C Ch 2018 *J. Mater. Process. Technol.* **251** 26-36 https://doi.org/10.1016/j.jmatprotec.2017.08.015.
[9] Song J L, Lin S B, Yang C L and Fan C L 2009 *J. Alloys Compd.* **488** 217-222 http://dx.doi.org/10.1016/j.jallcom.2009.08.084.
[10] Lv S X Jing X J, Huang Y X, Xu Y Q, Zheng C Q and Yang S Q *Sci. Technol. Weld Joi.* **17** 519–24 http://dx.doi.org/10.1179/1362171812Y.0000000041.
[11] Chen S, Li L, Chen Y and Huang J *J. Alloys Compd.* **509** 891-8 http://dx.doi.org/10.1016/j.jallcom.2010.09.125.

[12] Chen S, Li L, Chen Y and Liu D 2010 *Trans. Nonf. Met. Soc. Chi.* **20** 64-70 http://dx.doi.org/10.1016/S1003-6326(09)60098-4.
[13] Chen S Li L Chen Y Dai J Huang J 2011 *Mater. Design* **32** 4408–16 http://dx.doi.org/10.1016/j.matdes.2011.03.074.
[14] Casalino G Mortello M and Peyre P 2015 *J. Mater. Process. Tech.* **223** 139-49 https://doi.org/10.1016/j.jmatprotec.2015.04.003.
[15] Chen Y Chen S Li L 2010 *Mater Design* **31** 227–33 https://doi.org/10.1016/j.matdes.2009.06.029.
[16] Nagy M, Behúlová M and Pérez M R 2019 *IOP Conf. Ser. Mater. Sci. Eng.* **465** 012006.
[17] ESAB 2016 Welding Filler Metal Databook. Available online at https://www.esabna.com/shared/documents/litdownloads/weldingfmdb/us/2016/gen26883h_filler_metal_databook_us2016.pdf.
[18] Destructive tests on welds in metallic materials – Transverse tensile tests (ISO 4136:2012).
[19] Metallic materials – Method of test at room temperature (ISO 6892-1:2016).
[20] Metallic materials. Vickers hardness test. Part 1: Test method (ISO 6507-1:2005).
[21] Destructive tests on welds in metallic materials – Hardness testing – Part 2: Microhardness testingof welded joints (ISO 9015-2:2016).
[22] Destructive tests on welds in metallic materials – Etchants for macroscopic and microscopic examination (ISO/TR 16060:2003).
[23] Batalu D, Coşmeleată G and Aloman A 2006 *U P B Sci. Bull. B* **68** 4 77-90.
[24] Sambasiva R A, Madhusudhan R G and Satya P K 2011 *Mater. Sci. Technol.* **27** 1 65-70.
[25] Sahul M, Sahul M, Vyskoc M, Caplovic D and Pasak M 2017 *J. Mater. Eng. Perform.* **26** 1346-57 https://doi.org/10.1007/s11665-017-2529-6.
[26] Tomashchuk I, Sallamanda P, Cicala E, Peyre P and Grevey D 2015 *J. Mater. Process. Technol.* **217** 96–104 https://doi.org/10.1016/j.jmatprotec.2014.10.025.

IOP Conf. Series: Materials Science and Engineering 1050 (2021) 012011 doi:10.1088/1757-899X/1050/1/012011

Examination of crystal plasticity in cold drawn tube forming using FEM and Voronoi tessellation

M Necpal, M Martinkovič and M Kuruc

Slovak University of Technology in Bratislava, Faculty of Materials Science and Technology in Trnava, Ulica Jána Bottu 2781/25, 917 24 Trnava, Slovak Republic

martin.necpal@stuba.sk, maros.martinkovic@stuba.sk, marcel.kuruc@stuba.sk

Abstract. The analyses of the tube drawing process can be divided into two basics parts of investigations - macro-scale and micro-scale. Macro-scale analyses involve investigation of dimensional accuracy of the formed specimen, drawing forces, strain, stress, etc. The analyses of the tube drawing process in the macro-scale have been studied experimentally or theoretically using numerical methods for many years. For the final product of drawing process is also important to know the surface quality for example. For this purpose, it is important to know micro-scale material properties (spatial distribution of grain neighbourhood orientation etc.), and examine the context of micro forming material. There has been a trend in the last decade to make a link between macro-scale forming and micro-scale material properties and name it Multi-scale modelling. This research illustrates the possibility to use Voronoi tessellation grain microstructure model in DEFORM FEM software to predict grain plasticity. The result obtained from numerical simulation is compared with direct reconstruction, derivation of Voronoi tessellation, and measurement using the line and also with analytical theoretical method of strain distribution in tube draw process throw the die.

1. Introduction

The high quality tubes are widely used in various industrial products such as automotive industry pipes for brake liquid, aviation industry coolers, etc. Due to its specially usage, high dimensional accuracy, surface roughness and mechanical properties are required. Cold drawing process is used for manufacturing precision metal tubes and pipes. To avoid material fatigue during draw process it is important to understand material macro formability and also microstructural geometry of material. For this investigation, numerical simulation and computer analyses, usually FEM simulations are widely used.

By macro-scale numerical simulation, it is possible to analyse the material flow. This kind of simulations was applied for a better understanding of the drawing process by Bella [1], Necpal and Martinkovič [2]. Keshavarz [3] in experimental and computational study of Ni-Based superalloy illustrated the need for incorporating the critical microstructural features in plasticity models for understanding the role of microstructure on material behaviour. Zhang [4] investigated the inner surface roughness evolution of the high quality aluminium alloy AA6061 rectangle tube during cold drawing process. The micro-scale analysis with the 3D crystal-plasticity model were conducted based on the macro-scale FEA result. The macro and micro coupling analysis provide a new way to FEM analysis of the formed part and the plastic deformation of metal material in cold drawing process. This research illustrates FLOWNET functionality in DEFORM 2D software using Voronoi tessellation.

2. Numerical simulation

DEFORM 2D and the Lagrangian calculation has been applied for numerical simulation of cold drawing technology. The initial axisymmetric geometric configuration has been set.

2.1. Workpiece and tools of material modeling

During cold drawing process, the material temperatures remain lower and therefore material behaviour can be considered as temperature independent. In this case, the power law constitutive equation for definition of material plastic properties can be used (Equation 1)

$$\sigma = c\,\varepsilon^{n}\,\dot{\varepsilon}^{\,m} + y \qquad [\%] \qquad (1)$$

where ε represents the equivalent effective plastic strain, $\dot{\varepsilon}$ represents the effective plastic strain rate, and c, n, m and y are the material parameters determined by means of tensile tests. These parameters are listed in Table 1. The tube material was considered as plastic, the hardening is assumed to be isotropic, and the yield function type is set as von Mises. Elastic material properties can be neglected, the workpiece is set as a plastic.

Table 1. Mechanical properties of the stainless steel materials.

Symbol	Material parameter	Value
c	Material constant [-]	178
n	Strain exponent [-]	0.61
m	Strain rate exponent [-]	0.02
y	Initial yield value [MPa]	235.5

2.2. Tools geometry, properties and simulation condition

The drawing velocity was set 9 m.min^{-1} for all passes in experimental drawing and also as simulation boundary conditions. The forming tools (mandrel and die) were considered as rigid bodies. Tool and feedstock dimensions are given in Table 2. The friction model between the tools and the material was chosen to be a shear-type with the value of 0.08 [5]. The geometry of the tube was meshed. The eight elements are across the wall thickness of the tube. This mesh of workpiece is sufficient for required accuracy and it is not necessary to use the re-mesh procedure during the calculation. The mesh of tools did not need to be generated for simulation.

Table 2. Tool and feedstock dimensions.

Symbol	Material parameter	Unit	Value
OD	Outer diameter of the un-deformed tube	mm	28
WT	Wall thickness of the un-deformed tube	mm	4
α	Approach die angle	°	12
ID	Inner diameter of the die	mm	24
LDS	Length of the die bearing surface	mm	4
ODP	Outer diameter of the plug	mm	18
PT	Plug and tube contact length "measure after simulation"	mm	8.8

In Figure 1, the results of 2D FEM simulation in DEFORM software can be seen. Main maximum strain in longitudinal direction through a die is slowly increasing. Rapid increase of strain is in the contact area of drawn pipe with the die, which shown a graphical behaviour of the strain on the internal side of pipe.

3. Analytical solution

Calculation of tube cross-section (area) reduction is given by

$$\text{Reduction} = \frac{S_0 - S}{S_0} \times 100 \qquad [\%] \qquad (2)$$

where the S_0 [mm^2] is the tube cross section area before drawing and S [mm^2] is the tube cross section area after drawing. In this example, $S_0 = 301.59$ mm^2, $S = 236.40$ mm^2. According to Equation (2), the cross-section reduction is 21.6 %.

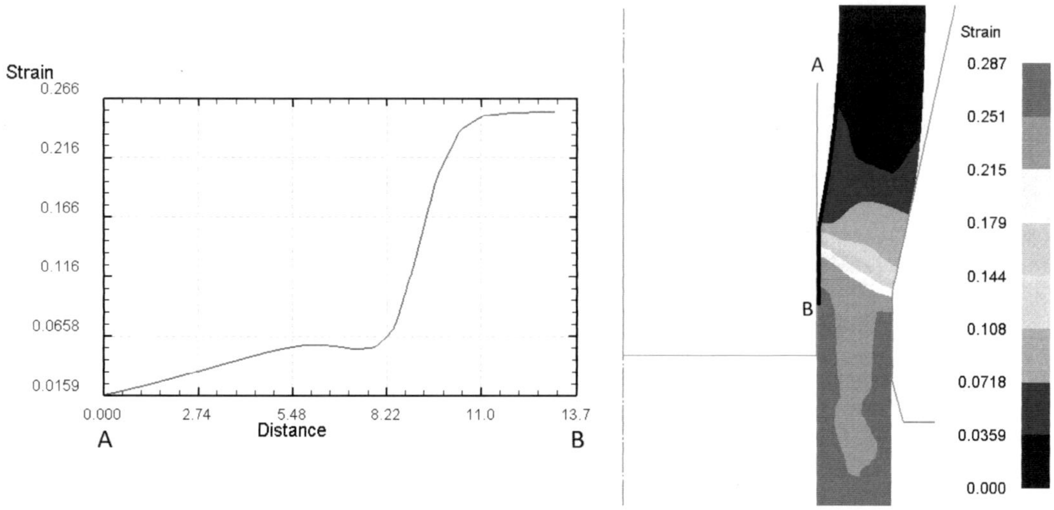

Figure 1. Results of 2D FEM simulation.

Deformation of a workpiece can be quantified by means of its dimensions before and after deformation. Let cylindrical bar has initial length l_0 and cross-section area S_0 and is formed by a reduction along its main axis. After deformation its length is l and cross-section S. The strain is defined as

$$\varepsilon = \frac{l - l_0}{l_0} \qquad [-] \qquad (3)$$

Forming law of Constant volume means that the volume of material before forming is equal to the volume of material after forming. Volume of drawing tube is defined as

$$V_0 = V = l_0\, S_0 = l\, S \qquad [-] \qquad (4)$$

Substituting Equation 3 to the Equation 4, it is possible to obtain Equation 5 to define the tension strain, which is primary strain of the tube drawing process.

$$\varepsilon = \frac{S_0 - S}{S_0} \qquad [-] \qquad (5)$$

In this case when the reduction is 21.6 %, the macroscopic tensile strain is $\varepsilon = 0.276$.

4. Investigation of plastic deformation by Voronoi tessellation boundary orientation

In this case, a combination of FEM simulation and FLOWNET pattern deformation is used to compare true strains. Tensile strain along the axis of the pipe is the main strain with maximum value in drawing process. Maximum main strain obtained by FEM simulation will be compared with strain obtained by boundary deformation degree of Voronoi tessellation.

4.1. *Polycrystalline morphology*

Mathematically, a Voronoi tessellation of an n-D space is collection of n-D entities that fills the space with no overlaps and no gaps. These entity are polyhedral and are formally defined as zones of influence of particular set of points, corresponding to their centres. Algorithms for the generation of a 2D computational mesh for the Voronoi cell are governed by the principles of Dirichlet tessellation in plane. Subdivision of a planar domain in the tessellation process is determined by a set of points belonging to the domain [6]. Several algorithms have been proposed in the literature to generate Voronoi tessellation or cells in plane. For our purposes, theMATLAB Voronoi standard functions [7] is used to compute the region of Voronoi diagram.

The generation procedure and result of Voronoi cells are illustrated in Figure 2. The first step is generation of random point, they represent the centres of Voronoi tessellation. The size of the region of Voronoi Tessellation is 4×2 mm, where 4 mm is the thickness of undeformed tube before forming operation.

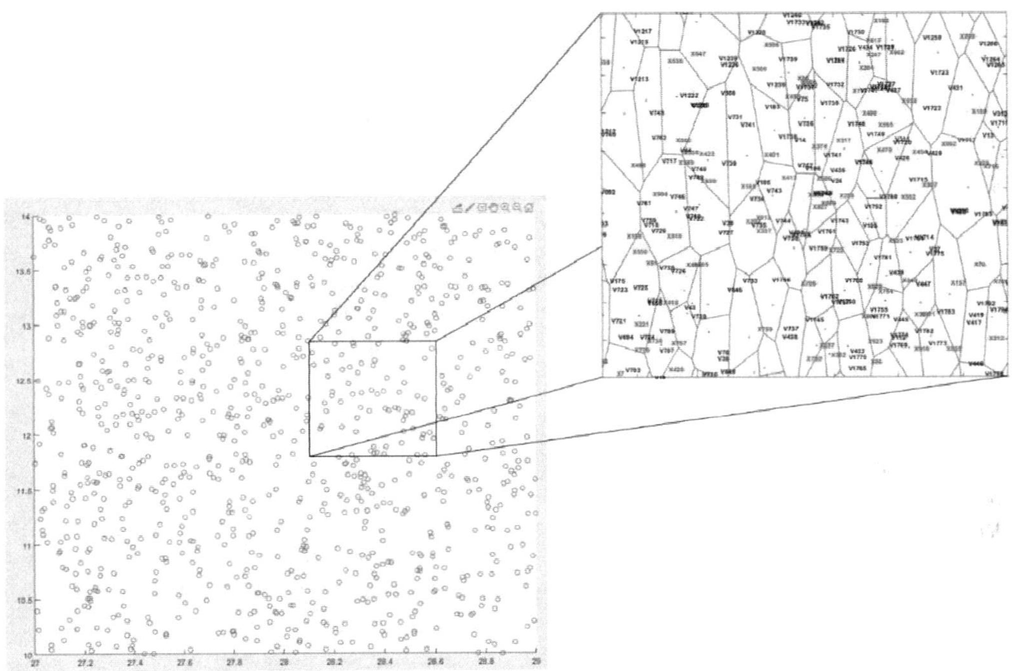

Figure 2. The generation procedure and result of Voronoi cells.

4.2. *FLOWNET DEFORM functionality*

The FLOWNET is a post-processing tool which allows the user to place some 2D pattern into the simulated region and let the simulation track the deformation of the pattern throughout the deformation [8], as shown Figure 3. As the FEM mesh deforms, so does the pattern, however, unlike the FEM mesh, the pattern remains intact also throughout the FEM mesh reconstruction (re-meshing). Thus, the FLOWNET is much like physically etching a pattern on a cross-section of the deformed workpiece. This is an excellent way in which to visualize any potential irregularities in the grain structure or to view potential surface defects such as folds. It should be noted that FLOWNET patterns can be generated for deforming objects. The start step list will contain all of the steps in the display window currently loaded by default.

Voronoi tessellation diagram generated by MATLAB has to be transformed as a user defined pattern file readable by DEFORM software. A pattern file consist of a list of point coordinates and a list of connectivity sets. The points are points of intersection within the FLOWNET.

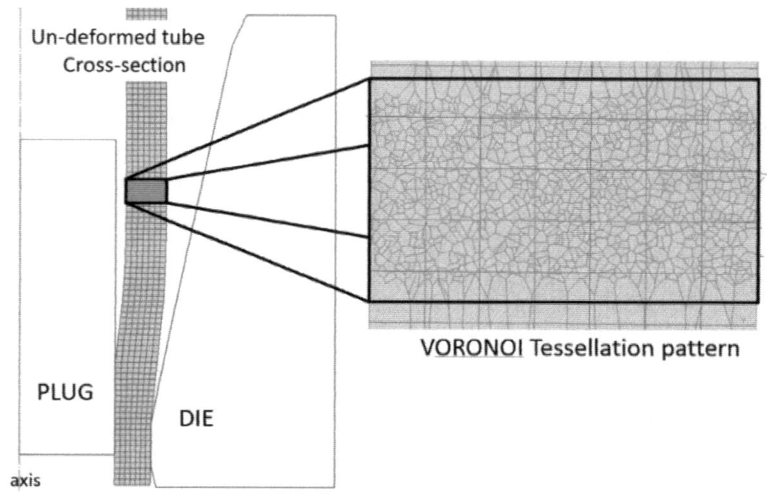

Figure 3. Voronoi tessellation pattern.

The strain in analysed area was obtained by measurement of degree of Voronoi tessellation boundaries orientation. On Voronoi tessellation area the test line were placed perpendicular and parallel to the boundaries orientation. From the specific number of parallel test lines intersections with tessellation boundaries $(P_L)_P$ and perpendicular lines ones $(P_L)_O$, the total specific surface area $(S_V)_{TOT}$ of cells and the planar orientation part of specific surface $(S_V)_{OR}$ of cells were estimated. From this values, the degree of orientation of cells boundaries O was estimated as ratio of $(S_V)_{OR}$ and $(S_V)_{TOT}$.

The result is proportional to Voronoi tessellation boundaries deformation degree and local plastic deformation, was estimated according to Equation 6 [9], where ε is the logarithmic strain.

$$\varepsilon = \ln\left(\frac{1+O\sqrt{2-O^2}}{1-O^2}\right)^{2/3} \qquad [-] \qquad (6)$$

Local plastic deformation, degree of orientation and estimated deformation according to Equation 6 in the zone of Voronoi tessellation pattern before and after the drawing process can be seen in Figure 4.

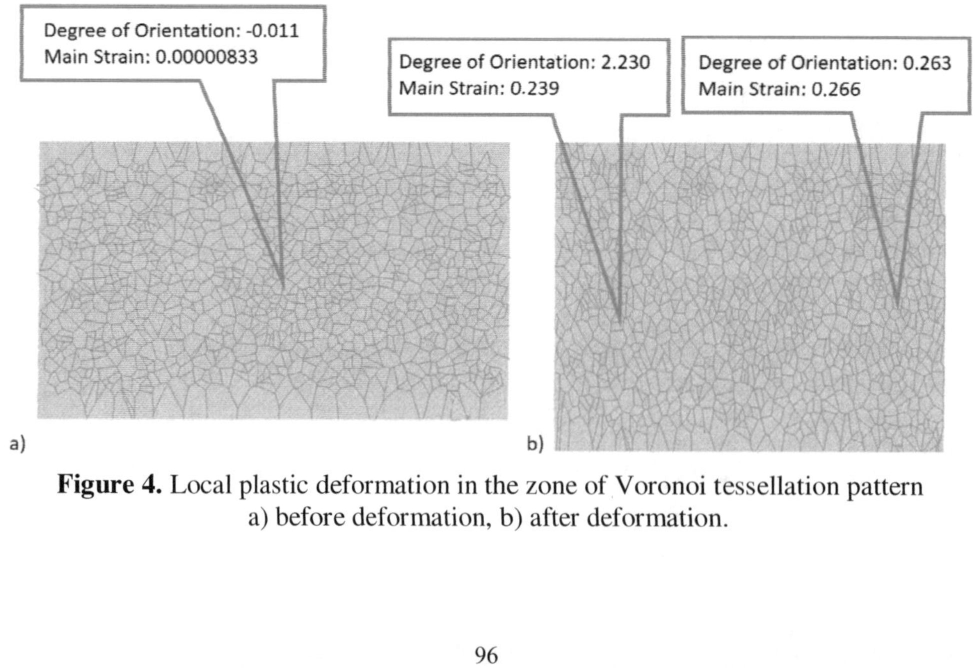

Figure 4. Local plastic deformation in the zone of Voronoi tessellation pattern a) before deformation, b) after deformation.

5. Conclusion

The FEM simulation of the deformation process of the formed material shows complex view of all parameters, not just strain, but also stresses or thermal influences, if they are included in the simulation. For confirmation of the correctness of the simulation model, it is generally required to check at least a part of the obtained results by different method. Analytical method is based on generally valid theoretical influences, it does not give exact information about specific material behaviour.

The use of the quantitative analysis of grain boundary orientation is giving exact view to material behaviour during its deformation. This work shows comparison of FEM analysis and the grain deformation simulation, which is represented as Voronoi tessellation. Qualitative analysis does not provide data in the specific point of material but within a certain volume. The results of individual analyses show that quantitative method of grain boundaries orientation can be used as supplement of computer simulation for confirmation of correctness of simulation model set up.

Acknowledgements

This work was supported by the Slovak Research and Development Agency under contract No. APVV-15-0319.

References

[1] Bella P, Ridzon M, Mojzis M and Parilak L 2017 The technology of cold drawing of seamless steel tubes using numerical simulation *Hutni-Wiadomości Hutnicze* **84-8** 356–58

[2] Necpal M and Martinkovič M 2019 Evaluation of material deformation during process of precise carbon steel tube cold draw forming *IOP Conf. Ser.: Mater. Sci. Eng.* **465** 012007

[3] Keshavarz S and Ghosh S 2013 Multi-Scale Crystal Plasticity Finite Element Model Approach to Modeling Nickel-Based Superalloys. *Acta Mater.* **61** 17 6549–61 https://doi.org/10.1016/j.actamat.2013.07.038

[4] Zhang L, Xu W, Long J and Lei Z 2015 Surface Roughening Analysis of Cold Drawn Tube Based on Macro–Micro Coupling Finite Element Method *J. Mater. Process. Technol.* **224** 189-99 https://doi.org/10.1016/j.jmatprotec.2015.05.009

[5] Necpal M, Martinkovič M and Václav Š 2018 Determination of the Coefficient of Friction Under Cold Tube Drawing Using FEM Simulation and Drawing Force Measurement *Research Papers: Faculty of Materials Science and Technology, Slovak University of Technology* **26** (42) 29-34 https://doi.org/10.2478/rput-2018-0003

[6] Ghosh S 2018 Micromechanical Analysis and Multi-Scale Modelling Using the Voronoi Cell Finite Element Method (Boca Raton: CRC Press) https://doi.org/10.1201/b10903

[7] MATLAB https://www.mathworks.com/help/matlab/voronoi-diagram.html?s_tid=CRUX_lftnav

[8] DEFORM™ v11.01 2014 Documentation, *Scientific Forming Technology Corporation*

[9] Martinkovič M and Václav Š 2013 Estimation of grain deformation of polycrystalline materials *Adv. Sci. Eng. Med.* **5** 585-8

Optical emission spectroscopy of radio frequency inductively coupled plasma for cold hydrogenation of nanoparticles

Z Remes[1], A Kromka[1], J Micova[2], B Rezek[3], A Poruba[4] and Hua-Shu Hsu[5]

[1] Institute of Physics Czech Academy of Sciences, Cukrovarnicka 10, 162 00 Praha 8, Czechia
[2] Institute of Chemistry Slovak Academy of Sciences, Dubravska cesta 9, 84538 Bratislava, Slovakia
[3] Faculty of Electrical Engineering, Czech Technical University in Prague, Technická 2, 166 27 Praha 6, Czechia
[4] SVCS Process Innovation s.r.o, Zámecká 133, 757 01 Valašské Meziříčí, Czechia
[5] Department of Applied Physics, National Pingtung University, Minsheng Rd. 4-18, 900 Pingtung City, Taiwan

remes@fzu.cz

Abstract. The radio frequency inductively coupled plasma (ICP) offers an alternative "cold" way to affect the size, composition, structure, and surface functionality of nanoparticles (NPs), such as metal oxide NPs, providing further adjustment of their physical and chemical properties. The ICP was monitored in-situ by optical emission spectroscopy (OES). In particular, hydrogen, oxygen, argon, and nitrogen plasma was studied. OES data show that despite the decrease of the optical emission intensity with increasing gas pressure, the concentration of atomic hydrogen increases with pressure and radio frequency power.

1. Introduction

Metal-oxide nanoparticles (NPs) with their high surface-to-volume ratio and related size effects are currently studied for energy conversion, photocatalytic waste water cleaning, electrochemical energy storage or sensing applications [1]. For example, zinc oxide (ZnO) is a versatile, environmentally friendly, functional semiconducting material with unique optical properties and variety of nanostructures such as nanorods, nanowires, nanobelts, nanotubes, nanoribbons or hedgehog-like NPs [2]. A low-cost technology for mass-production of ZnO NPs is a hydrothermal chemical synthesis [3]. Recently, we have shown that the UV illumination [4] and the age of precursors [5] plays a significant role in defects formation in hydrothermally grown ZnO NPs.

The interstitial hydrogen acts always as a donor in ZnO [6]. It has been shown that the high temperature annealing of ZnO in H_2 atmosphere passivates defects and acts as reducing agent creating oxygen vacancies [7]. To avoid high temperature, inductively coupled plasma (ICP) offers an alternative way of cold hydrogenation. Plasma effects the size, composition, structure, surface, and functions of NPs providing further adjustment of their physical and chemical properties [8]. We have shown that the plasma hydrogenation is an effective low temperature way to suppress defects in ZnO NPs leading to significant enhancement of an exciton-related emission band [9]. The optical emission spectroscopy (OES) provides in-situ information about chemical and physical processes that occur in plasma. Using OES we investigate in this paper the effect of pressure and radio frequency (rf) power on dissociation of H_2 molecules.

2. Experimental

2.1. Inductively coupled plasma (ICP)

The plasma hydrogenation of NPs is done in a prototype ICP reactor currently being developed in a cooperation with SVCS Process Innovation, s. r. o., Valašské Meziříčí, Czech Republic. ICP operates at frequency 13.56 MHz and rf discharge power up to 300 W using argon (99.998%), hydrogen (99.999%), oxygen (99.995%), and nitrogen (99.999%) process gasses. Prior any plasma treatment, the chamber, valves, flowmeters, gauges, and all gas inlet and outlet tubes are evacuated with the Agilent Triscroll 300 dry vacuum pump protected with the ISO Flange Vacuum Filter (CSL-843-NW25) to residual pressure below 0.1 Pa and flushed 5 min by 50 sccm flow of process gases to reduce residual gas contamination. A powder sample is placed in a grounded cradle-like rotary stainless steel holder and stirred during the plasma treatment by swinging the cradle to achieve a good homogeneity, see H_2 plasma in figure 1.

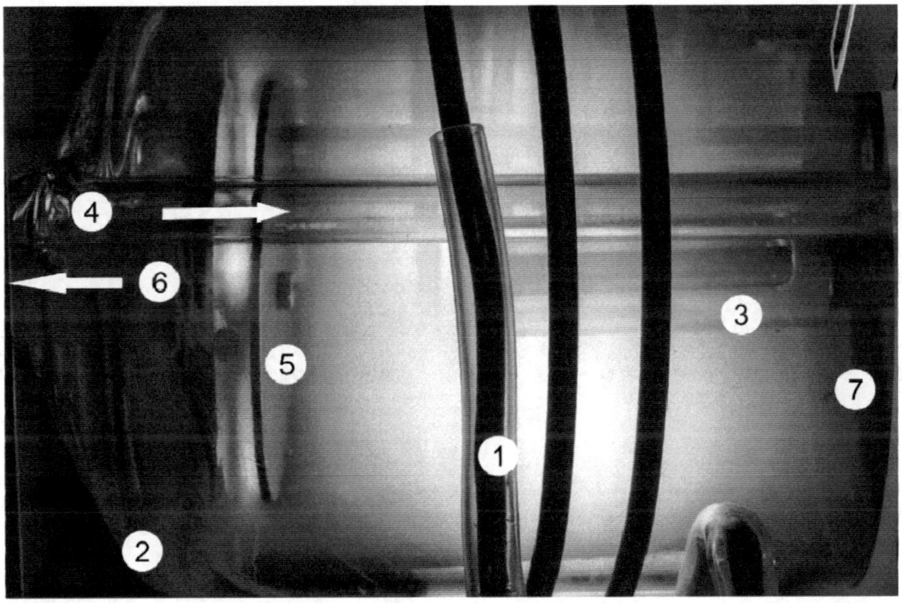

Figure 1. The ICP reactor: 1 - external copper wire coil, 2 - quartz chamber, 3 - grounded stainless steel sample holder ("a cradle") with a stainless steel rotating vacuum feed-through, 4 - gas inlet quartz tube, 5 - plasma shield in front of a gas outlet, 6 - gas outlet, 7 – door.

2.2. Optical emission spectroscopy (OES)

Optical emission spectra (OES) are measured in 400–1000 nm spectral range by spectrally calibrated fibre coupled TE cooled CCD spectrometer (B&W Tek BTC112E) equipped with a solarisation resistant optical fibber and quartz lens focused in the middle of the quartz chamber above the grounded stainless steel sample holder. The integration time varies from 10 ms to 10 s depending on optical emission intensity, each spectrum is averaged 10 times and divided by the integration time to be comparable with other spectra. Dark spectra are measured without plasma and subtracted. The whole setup was spectrally calibrated with Oriel #63358 quartz tungsten halogen lamp and a white optical diffuser placed inside the quartz chamber.

3. Results and discussion

Each gas has a specific plasma colour. Therefore, OES can be used as a fingerprint for gas identification, see figure 2. Hydrogen plasma has a magenta colour. The red 656 nm (H_α), blue 486 nm (H_β), and violet 434 nm (H_γ) spectral lines in the visible part of the emission spectrum are easily recognizable as Balmer series of atomic hydrogen H [10]. Argon has a very large electron impact ionization cross section, high degree of ionization in plasma and the most intense emission peaks in the near infrared region at 764, 810, 841, 842, and 912 nm. Oxygen plasma has a pale white colour due to the lack of strong emission lines in the visible region. The characteristic emission lines of atomic oxygen appear in near infrared region at 777 and 845 nm. The dissociation energy of N_2 molecule is very high (9.8 eV) and therefore the dissociation ratio of N_2 is low in the ICP [11]. Nitrogen plasma has an orange colour with typical diatomic emission bands [12]. OES may be used for identification of possible plasma contamination, such as leakage of N_2 from the atmosphere into the reactor, which may be the case of H plasma in figure 2.

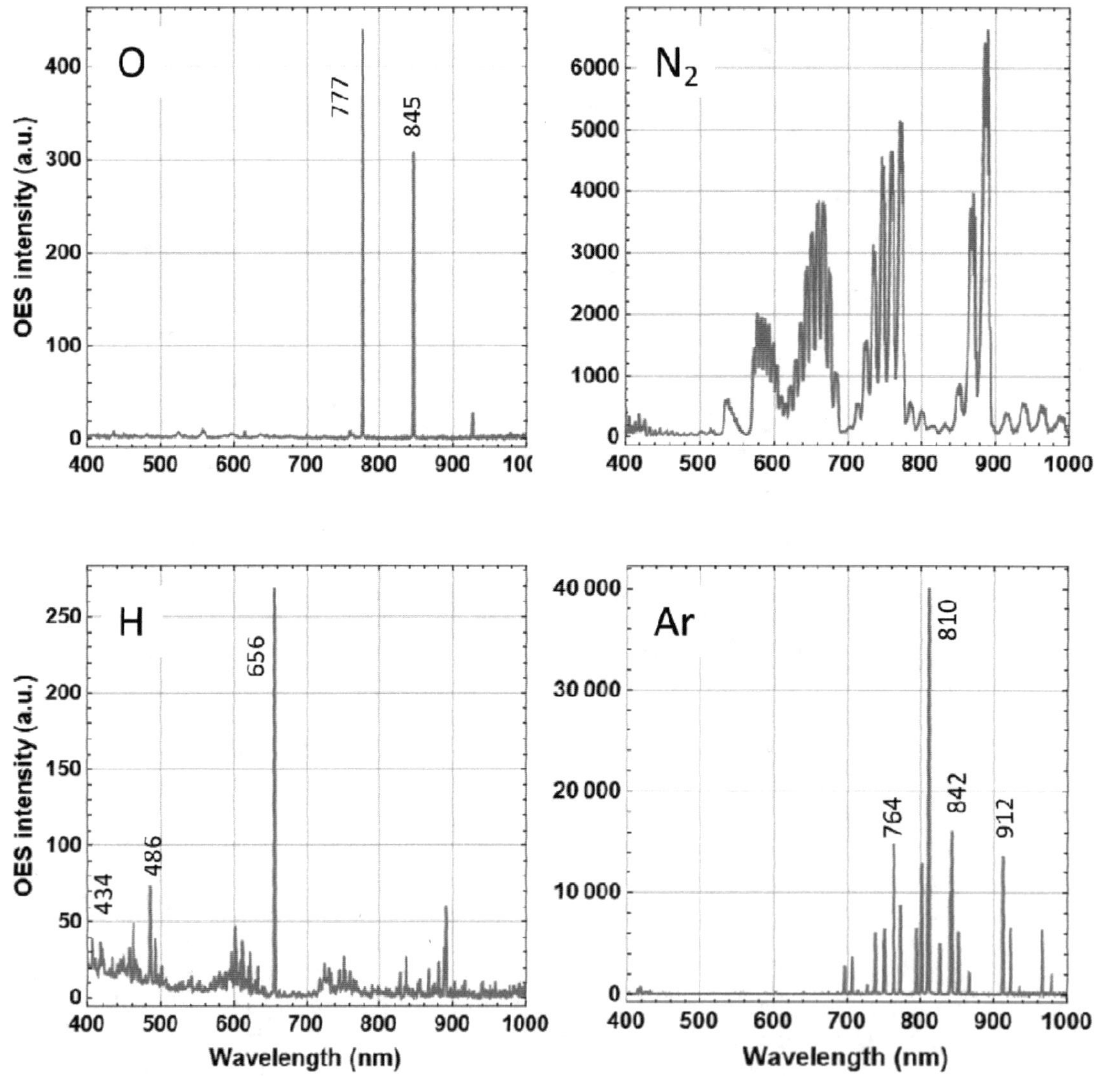

Figure 2. OES spectra of O, N_2, H, and Ar plasma measured at the same pressure 20 Pa and rf power 200 W, except of Ar were rf power had to be reduced to 50 W to avoid saturation of OES spectrometer.

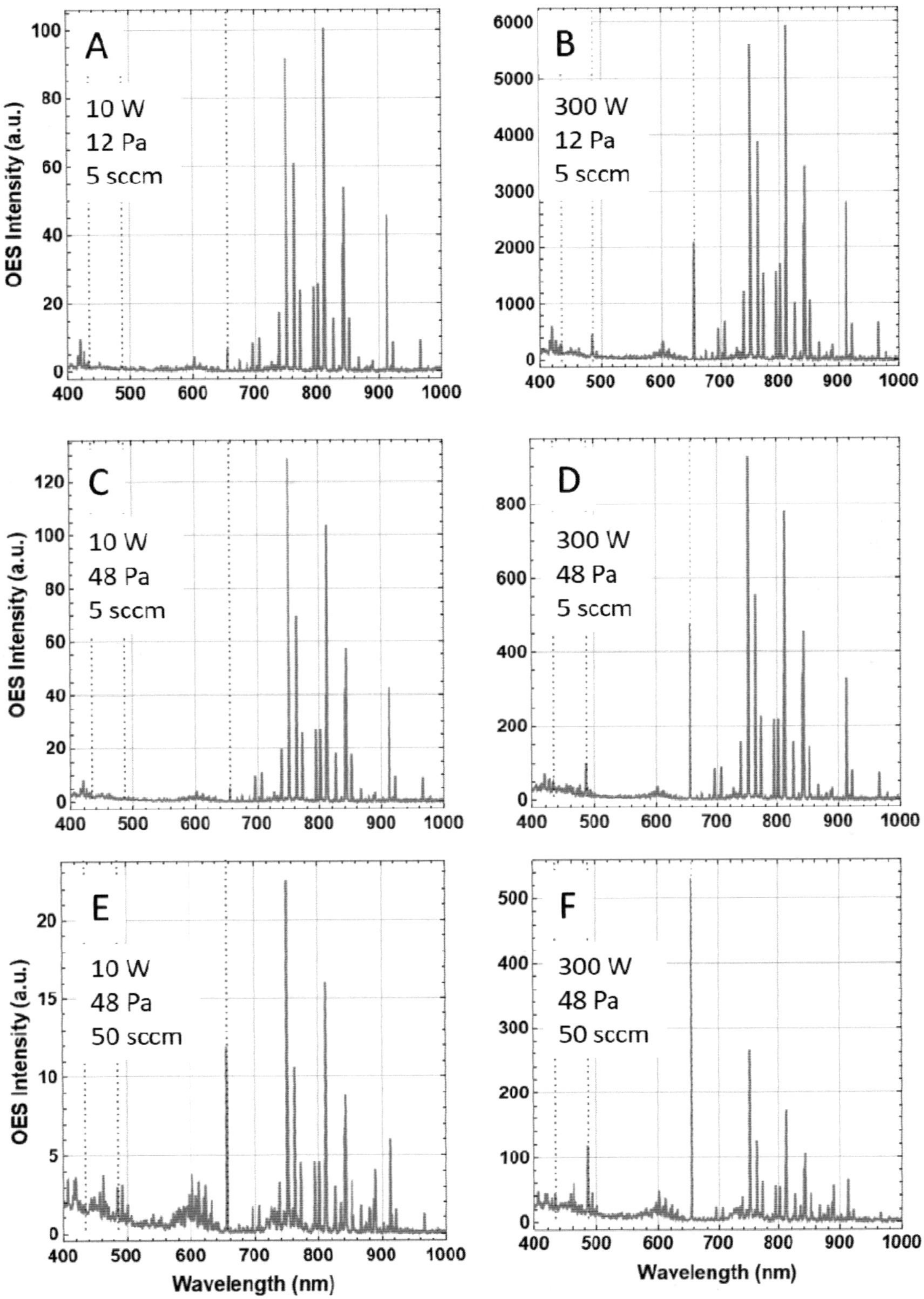

Figure 3. OES spectra (relative intensities) of H_2 plasma diluted with 2 sccm Ar flow. The rf power, total pressure, and H_2 flow rates are indicated in each graph. The H-related OES peaks at 656, 486, and 434 nm are indicated with dashed vertical lines.

The ionization energies are 15.76 eV for Ar, 13.60 eV for atomic hydrogen, 15.42 eV for H_2 molecule and the dissociation energy for hydrogen molecule H_2 is 4.48 eV [13]. The main mechanism for dissociation of H_2 molecules into H atoms is the impact excitation due to inelastic collisions of H_2 molecules with hot electrons *e*, see equation (1) [8].

$$e_{\mathrm{hot}} + H_2 \rightarrow 2H + e_{\mathrm{cold}} \tag{1}$$

The addition of Ar into hydrogen plasma induces several effects related to the high degree of Ar ionization. Relatively heavy Ar^+ cations cause heating of the sample by ion bombardment. The ionization increases the density of plasma electrons and related electrical conductivity of plasma. On the other hand, hot electrons are cooled via inelastic collisions with heavy Ar atoms, see equation (2).

$$e_{\mathrm{hot}} + Ar \rightarrow Ar^* + e_{\mathrm{cold}} \tag{2}$$

The other effect is a series of resonant energy transfer channels from excited Ar^* atoms to H atoms, see equation (3) [14].

$$Ar^* + H \leftrightarrow Ar + H^* \tag{3}$$

For the purpose of our OES study, the constant 2 sccm Ar flow was mixed with the H_2 flow that varied between 5–50 sccm. The total gas pressure increased from 12 Pa for 5 sccm H_2 flow to 48 Pa for 50 sccm H_2 flow. The non-linearity of total pressure with gas flow was caused by low pumping speed at low pressure that varies with species causing difficulty in estimation of partial pressures in a mixture of gases with dissimilar molecular weights. Figure 3 shows that the OES intensity increases with rf power but the increased pressure significantly reduces OES intensity of both species. The OES intensity is quenched at higher pressure due to the increased collision frequency and the non-radiative transition of electrons from excited states. Therefore, the degree of H_2 dissociation cannot be directly evaluated using H emission line intensities alone.

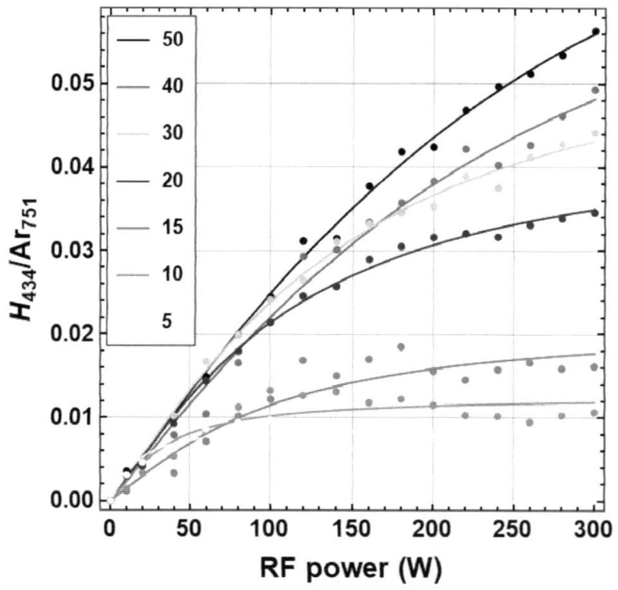

Figure 4. The ratio of integral intensities of the atomic H and Ar lines at 434 and 751 nm bands as a function of rf power and H_2 flow rates 5, 10, 15, 20, 30, 40, and 50 sccm with a constant Ar flow (2 sccm) corresponding to the total gas pressure 12, 22, 24, 27, 34, 41, and 48 Pa. Smooth curves were added as eye guides.

Actinometry had been proposed as a method for evaluating the relative degree of molecule dissociation independently of optical emission quenching effects at higher pressure [15]. In our case, the relative concentration of atomic H radicals was calculated from the integral ratio of OES intensities of H_{434} and Ar_{751} emission lines. The lines were chosen because of the comparable excitation threshold energies of the excited states (13.06 eV for H_{434} and 13.48 eV for Ar_{751}) relative to their ground state [16]. Prior the intensity integration, the broad band background emission was digitally subtracted from emission spectra. Figure 4 shows that the H_{434} to Ar_{751} intensity ratio and therefore the degree of H_2 dissociation increases sub-linearly at low rf power and saturates at higher rf power. These results suggest that the hydrogen molecular dissociation is more efficient at higher pressures when using higher rf power.

4. Conclusions

The optical emission spectroscopy (OES) provides information about chemical and physical processes occurring in plasma. OES is also useful to control of possible contamination of plasma. Actinometry is a relatively simple procedure to evaluate degree of molecular dissociation, designed to overcome the effect of plasma excitation efficiency from the OES data assuming that the optical cross sections of the inert gas and the active species are equal. OES spectra confirmed that the concentration of atomic hydrogen increases in ICP with rf power and gas pressure and saturates with a power threshold increasing with total pressure. Therefore, the ICP is an effective method of generating atomic hydrogen by dissociation of H_2 without exposing a powder sample to high temperature The plasma hydrogenation of NPs at room temperature will provide further adjustment of their properties avoiding thermally activated processes such as thermal expansion and diffusion.

Acknowledgements
This work was supported by the Czech Science Foundation project 19-02858J, by the Operational Programme Research, Development and Education (project SOLID21 - CZ.02.1.01/0.0/0.0/16_019/0000760), by the Scientific Grant Agency (VEGA, Slovak Academy of Sciences, grant No. 2/0157/20), and by the Ministry of Science and Technology of Taiwan under grant Nos. 108-2923-M-153 -002 -MY3. Special thanks to Yu Ying Chang for technical help with ICP setup.

References

[1] Jelinek R 2015 *Nanoparticles* (Berlin ; Boston: Walter de Gruyter GmbH & Co., KG)
[2] Klingshirn C F 2010 *Zinc Oxide: from fundamental properties towards novel applications* (Heidelberg ; London: Springer)
[3] Mičová J, Buryi M, Šimek D, Drahokoupil J, Neykova N, Chang Y-Y, Remeš Z, Pop-Georgievski O, Svoboda J and Im C 2018 Synthesis of zinc oxide nanostructures and comparison of their crystal quality *Applied Surface Science* **461** 190–5
[4] Neykova N, Chang Y-Y, Buryi M, Davydova M, Kucerkova R, Simek D, Remes Z and Pop-Georgievski O 2019 Study of ZnO nanorods grown under UV irradiation *Applied Surface Science* **472** 105–11
[5] Buryi M, Babin V, Chang Y-Y, Remeš Z, Mičová J and Šimek D 2020 Influence of precursor age on defect states in ZnO nanorods *Applied Surface Science* **525** 146448
[6] Van de Walle C G and Neugebauer J 2003 Universal alignment of hydrogen levels in semiconductors, insulators and solutions *Nature* **423** 626–8
[7] Procházková L, Gbur T, Čuba V, Jarý V and Nikl M 2015 Fabrication of highly efficient ZnO nanoscintillators *Optical Materials* **47** 67–71
[8] Grill A 1994 *Cold plasma in material fabrication: from fundamentals to applications* (New York: the Institute of Electrical and Electronics Engineers, Inc.)
[9] Chang Y-Y, Remes Z and Micova J 2020 Mass production of hydrogenated ZnO nanorods *NANOCON 2019 - Conference Proceedings* 11th International Conference on Nanomaterials

- Research and Application, 16-18 October 2019 (Hotel Voronez I, Brno, Czech Republic: Tanger Ltd., Ostrava) pp 221–5

[10] Kramida A and Ralchenko Y 1999 NIST atomic spectra database, NIST standard reference database 78

[11] Herzberg G 1928 The dissociation energy of nitrogen *Nature* **122** 505–6

[12] Lofthus A and Krupenie P H 1977 The spectrum of molecular nitrogen *Journal of Physical and Chemical Reference Data* **6** 113–307

[13] Corrigan S J B 1965 Dissociation of molecular hydrogen by electron impact *The Journal of Chemical Physics* **43** 4381–6

[14] Krogh O, Wicker T and Chapman B 1986 Summary abstract: Collisional energy transfer in noble gas/hydrogen plasmas *Journal of Vacuum Science & Technology* A: *Vacuum, Surfaces, and Films* **4** 1796–7

[15] Coburn J W and Chen M 1980 Optical emission spectroscopy of reactive plasmas: A method for correlating emission intensities to reactive particle density *Journal of Applied Physics* **51** 3134–6

[16] Mortet V, Hubicka Z, Vorlicek V, Jurek K, Rosa J and Vanecek M 2004 Diamond growth by microwave plasma enhanced chemical vapour deposition: Optical emission characterisation and effect argon addition *physica status solidi* (a) **201** 2425–31

IOP Conf. Series: Materials Science and Engineering 1050 (2021) 012013 doi:10.1088/1757-899X/1050/1/012013

Bactericidal effect of zinc oxide nanoparticles on Gram-positive and Gram-negative strains in reverse spin bioreactor

D Rutherford[1], J Jíra[1], K Kolářová[2], Z Remeš[2] and B Rezek[1]

[1] Faculty of Electrical Engineering, CTU, Technická 2, 16227 Prague, Czech Republic
[2] Institute of Physics, CAS, Cukrovarnická 10, 16200 Prague, Czech Republic

david.rutherford@fel.cvut.cz

Abstract. We present an experimental study investigating the efficacy of zinc oxide nanoparticle use for bacterial decontamination of water using reverse spin technology. The number of viable bacteria reduced with treatment time and the rate of inactivation was dependent on zinc oxide concentration. Gram-positive *S. aureus* bacteria were more resistant to zinc oxide nanoparticles compared to Gram-negative *E. coli*, however both strains were completely eradicated after 4 hours using a concentration of 10 µg/mL. Nanoparticles did not inhibit growth of bacteria when added to an agar surface, neither when in liquid before bacteria inoculation nor when loaded onto filter paper disks after bacteria inoculation. These results emphasise the importance of efficient reactor mixing to enhance interaction capability for ZnO use in water decontamination.

1. Introduction

Water-borne diseases which arise from consumption of water of poor quality contribute to over 80% of reported illnesses in the developing world and it's estimated that around 4 billion people worldwide lack access to sanitized water [1]. The presence of pathogenic microorganisms, such as bacteria and viruses, are thought to be the main causative agents of water-borne diseases and disinfection strategies are necessary to provide people with an adequate supply of drinkable water. The ever-growing human population puts increasing pressure on a finite water resource and there is a need for innovative solutions for water decontamination that are efficient, low-cost and environmentally friendly. There are already many established methods such as chlorination, (electro)oxidation, illumination by ultraviolet light, or photocatalytic decomposition. All these methods have some limitations or drawbacks.

Nanotechnology has shown great potential for inactivating bacteria in water through the application of semi-conductive photocatalytic materials with low toxicity such as titanium dioxide and zinc oxide. The fundamental mechanism entails the formation of electron-hole pairs upon photon absorption which results in the initiation of a complex cascade of chemical reactions and the formation of reactive oxygen species (ROS) e.g. OH, O_2^-, H_2O_2 [2]. These radicals can rapidly oxidize biomolecules such as DNA, protein and lipid to cause irreversible cell damage and lead to cell death [3]. Recently, nanomaterials with low toxicity that retain a strong antibacterial effect without the need for additional illumination have become a viable alternative to conventional materials used in photocatalytic disinfection of water by eliminating the requirement for additional light source thereby reducing energy consumption without any adverse effect on decontamination efficacy [4].

Efficient mixing of the contaminated water with nanoparticles increases treatment efficacy. Reverse spin technology is a mechanically-driven mixing technique that rotates the cell-nanoparticle suspension about its axis with a change of direction of rotation motion [5]. This non-invasive technique has low energy consumption and can operate with or without additional illumination, making the apparatus ideal for investigating the antibacterial effects of nanoparticles.

2. Materials and Methods

2.1. *Zinc oxide nanoparticle preparation and characterisation*

A stock suspension of 0.2 mg/mL zinc oxide nanoparticles (ZnO NP, 50 nm nominal size, Sigma) was prepared using deionised water (dH_2O), autoclaved (121 °C for 15 min) then sonicated before use (37 kHz, 100 W, 30 min). Any dilutions made from the stock solutions were made using dH_2O. Material was characterized using SEM (Tescan MAIA3) and UV-vis spectroscopy (Biotek Epoch 2 96-well microplate spectrophotometer).

2.2. *Bacteria preparation*

Bacteria strains were purchased from Czech collection of microorganisms, one representative of Gram-positive bacteria (*S. aureus*) and one representative of Gram-negative bacteria (*E. coli*). Frozen bacteria stock (in Mueller Hinton broth, MHB, and glycerol at -20 °C) were allowed to thaw at +4 °C then 100 μL was added to MH agar and spread evenly across the surface and incubated for 18-24 h at 37 °C. From the resultant growth, one colony was removed and added to 10 mL MHB and placed on an orbital shaker (200 rpm) positioned inside an incubator set at 37 °C for 18-24 h. The bacteria suspension was adjusted to 0.5 McFarlands density in dH_2O, then subsequently diluted by 1:1000 to achieve an approximate initial cell concentration of 1×10^5 colony forming units per millilitre (cfu/mL).

2.3. *Antibacterial tests*

Standardised microbiological tests were used to assess the antibacterial capability of ZnO. Broth dilution assay [6] was employed to determine the number of bacteria that remained viable after exposure to ZnO. Suspensions of bacteria and ZnO NP in dH_2O were added (1:1) and subjected to reverse spin technology (200 rpm; 5 s per spin) and samples were removed every 2 h (see figure 1).

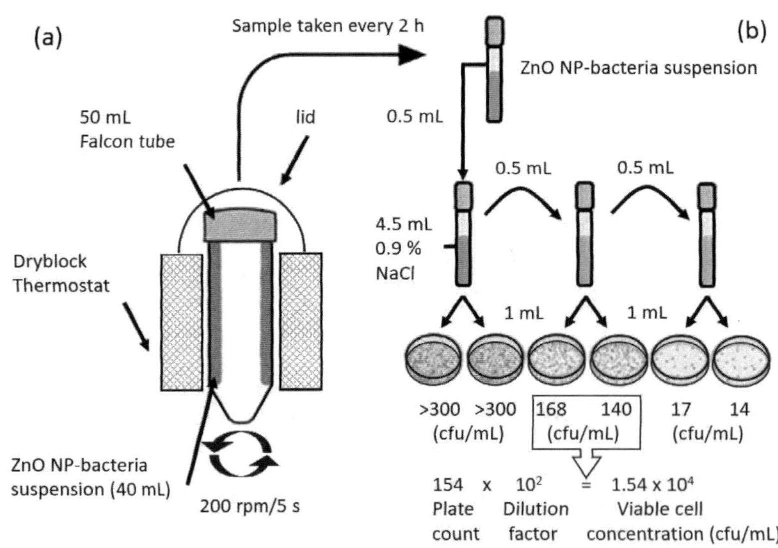

Figure 1. Schematic representations of (a) reverse spin technology used to expose bacteria to ZnO NP and (b) the broth dilution technique [6] to determine the number viable bacteria cells from a sample.

The disk diffusion assay [7] was also used to probe the antibacterial effect when bacteria were present on an agar surface. For comparison, ZnO NP were added to the agar surface prior to the inoculation of a known concentration of bacteria and the resulting colonies that grew were counted.

3. Results

The commercial ZnO used in this study was characterised using SEM and UV-vis spectroscopy. Material morphology was probed by SEM in both in-beam secondary electron and back-scattered electron modes (see figure 2(a) and 2(b), respectively). The images revealed that the primary particle shapes were grain-like, although the supplier stated the shape to be spherical. Also, we observed large agglomerates of primary particles up to 1 μm in diameter, which is approximately the same size as a bacterium. The bacteria-ZnO NP interaction was studied in ambient light conditions. Thus, UV-vis spectroscopy was applied to investigate the light absorbance capabilities of the material. Figure 2(c) shows the light absorbance spectral profile obtained from ZnO NP in dH_2O with an absorbance peak at 363 nm which is typical for ZnO.

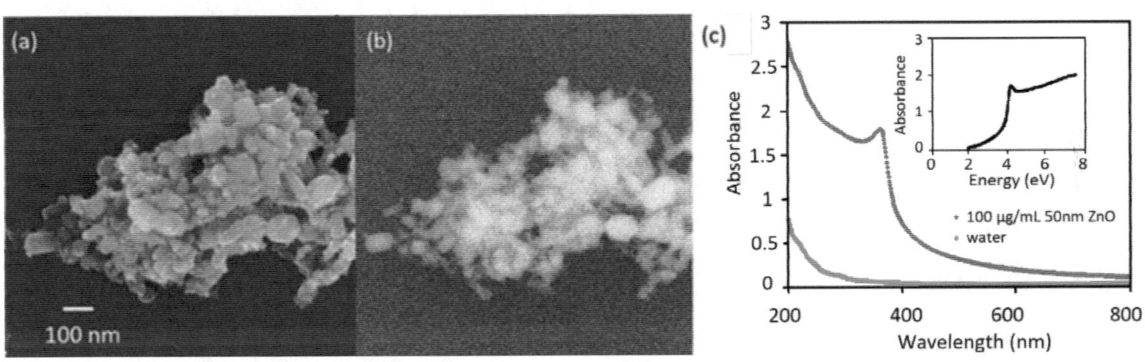

Figure 2. Scanning electron micrograph of ZnO NP drop-casted onto silicon substrate: (a) in-beam secondary electron (SE) characterisation of morphology and (b) back-scattered electron (BSE) showing primary particle-particle contact. (c) UV-vis absorbance of ZnO NP as a function of wavelength. Graph inset shows blank-corrected absorbance spectrum against photon energy.

We employed standardised microbiological techniques to characterize the antibacterial potential of commercially available ZnO NP. The first of these was the "broth dilution" technique (see the process scheme in figure 1(b)), which measured the number of bacteria that retained the ability to grow after exposure to ZnO NP using reverse spin technology. The results are shown in figure 3 for both bacteria strains and two concentrations of ZnO. The reference samples had only bacteria in water and no ZnO, and as expected there was no reduction observed under these conditions (see figure 3). It is clear from figure 3 that the antibacterial effect of ZnO is dependent on exposure time, ZnO concentration, and cell type. Complete eradication of both bacteria strains was achieved after 4 hours at a concentration of 10 μg/mL. The equivalent exposure time using 1 μg/mL only achieved between 1.5-2 orders of magnitude reduction in viable cells. At each sample point, the number of viable Gram-positive *S. aureus* was greater than the number of Gram-negative *E. coli*.

There was no antibacterial effect observed when ZnO NP was added to the agar surface prior to bacteria addition (see figure 4(b)). Similar results were observed when ZnO NP were added to filter paper disks for the concentration range used in this study (10 – 1000 μg/mL, see figure 4(c).

4. Discussion

Our observations indicate that only under certain conditions does ZnO exert an antibacterial effect. The constant mixing of the bacteria-nanoparticle suspension by reversing the direction of rotation enhanced the interaction capability and the antibacterial effect of ZnO NP. Yet, when ZnO NP were added to filter paper disks and placed upon bacteria, there was no observed antibacterial effect. This

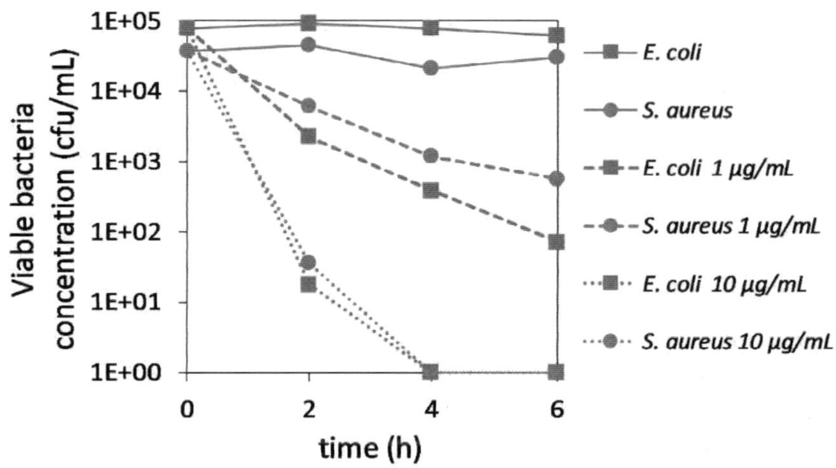

Figure 3. Effect on viable cell concentration as a function of exposure to two concentrations of ZnO. Solid lines are reference samples and contain only bacteria in water i.e. no ZnO (red – *E. coli* and blue – *S. aureus*). Dashed lines represent bacteria exposed to ZnO with stated concentration.

static type interaction is vastly different to the interaction mechanism involved in reverse spin technology. A second type of static interaction, ZnO NP added to agar before bacteria, also did not have any bactericidal effect. These observations confirm the importance of efficient sample mixing in the antibacterial effect of ZnO.

When the interaction between bacteria and ZnO NP was enhanced using the reverse spin technology, the "broth dilution" assay revealed three distinct ways that ZnO acted upon bacteria to render them non-viable. The first was related to the exposure time, and the longer the cells were exposed to ZnO the greater the antibacterial effect. Commercial ZnO with an average particle size of 30 nm at a concentration of 10 mg/mL was able to reduce pathogenic *E. coli* (0157:H7) by 1 order of magnitude after 4 hours that increased to 2 orders of magnitude by 8 hours exposure time [8]. The interaction between bacteria and ZnO NP occurred in liquid, but it is not evident whether the

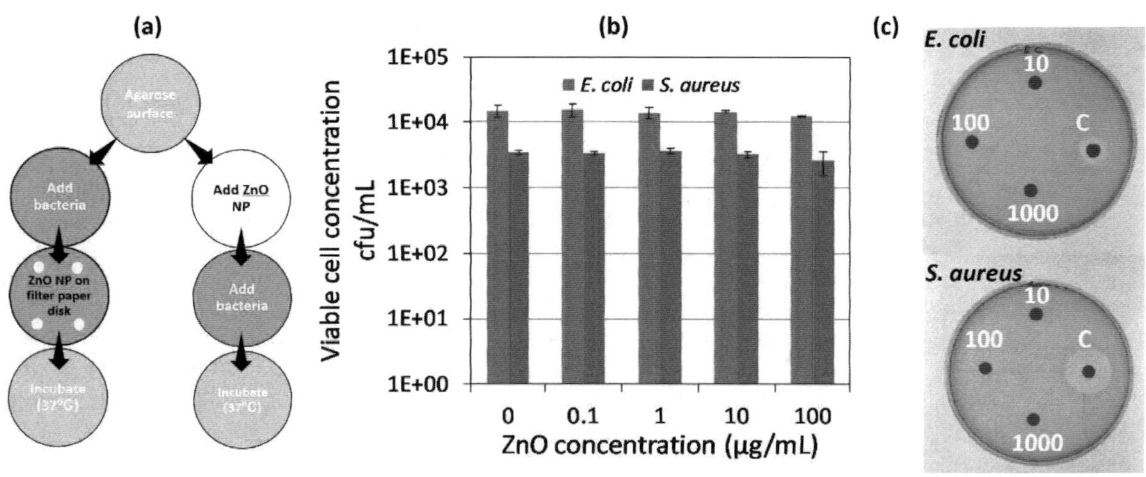

Figure 4. (a) Schematic diagram of the two antibacterial tests. (b) The concentration of viable cells after exposure to different concentrations of ZnO NP on agar surface. (c) Disk diffusion assay: Mueller Hinton agar plates with *E. coli* (top) and *S. aureus* (bottom). Numbers denote concentration of ZnO on disk in µg/mL. C = positive control (50 µL of 0.1 M H_2O_2).

suspension was mixed throughout exposure. We observed a 5 orders of magnitude reduction of non-pathogenic *E. coli* with a considerably lower concentration of 10 µg/mL 50 nm ZnO.

The second way was associated with ZnO concentration, and the positive correlation with bactericidal effect as shown by greater reduction in viable cells observed with the higher concentration of ZnO. We achieved complete inactivation of both Gram-positive *S. aureus* and Gram-negative *E. coli* within 4 hours using 10 µg/mL, yet for the same treatment time viable cells remained after exposure to 1 µg/mL. Navale *et al.*, incubated synthesised ZnO NP with *S. aureus* and mixed the suspension at 200 rpm for 4 hours in one direction i.e. without reverse spin, and a complete loss of viability was achieved using a 10-fold greater ZnO NP concentration (100 µg/mL) [9].

Finally, there appeared to be differences in antibacterial effect dependant on the type of bacteria cell being exposed, since Gram-negative *E. coli* were reduced to a greater extent than Gram-positive *S. aureus* at every sample time point. This can be explained by differences in the outer cell wall structure of the two bacteria types. *S. aureus* has a thick layer of peptidoglycan (~30 nm, [10]) which plays an important structural role and provides rigidity, whereas the peptidoglycan layer is much thinner in *E. coli* (7-9 nm, [11]) and would be more susceptible to mechanical damage.

The results from our investigation lead us to believe that the primary mechanism by which ZnO NP exert an antibacterial effect using reverse spin technology involves direct interaction of ZnO NP with the bacteria cell surface which can cause damage and lead to cell death. The constant mixing prevents nanoparticle agglomeration and enhances the efficiency of ZnO NP. Other factors such as the chemical reactions induced by electron-hole pair formation on the nanoparticle surface would not be dominant due a lack of light irradiation since the bacteria-nanoparticle suspension was shielded during exposure. Whilst reactive oxygen species can be generated from zinc oxide surface without illumination [12], the concentrations produced would be too low to be responsible for the observed bactericidal effect [13].

5. Conclusion

We achieved complete eradication of both model Gram-positive and Gram-negative bacteria in water using commercially available nano-sized ZnO without additional illumination. When the bacteria-nanoparticle suspension was constantly mixed using a reverse spin bioreactor, no viable cells remained after only 4 hours treatment time using a relatively low concentration of ZnO. However, no antibacterial effect was observed when the bacteria and nanoparticle suspensions were not sufficiently mixed together, either when ZnO was present on the agar surface nor on filter paper disks even for much higher concentrations. The reduction in viable bacteria using ZnO and reverse spin technology was achieved without the need for additional illumination sources, therefore reducing the potential cost and energy consumption of the treatment compared to conventional photocatalytic reactors which use UV light and TiO_2.

Acknowledgements

The presented work was supported financially by the project GACR 19-02858J. We also acknowledge use of CzechNanoLab Research Infrastructure supported by MEYS CR (LM2018110).

References

[1] World Health Organisation 2003 Emerging issues in water and infectious disease

[2] Ahmed S N and Haider W 2018 Heterogeneous photocatalysis and its potential applications in water and wastewater treatment: a review *Nanotechnology* **29** 342001

[3] Winterbourn C C 2008 Reconciling the chemistry and biology of reactive oxygen species *Nat Chem Biol* **4** 278–86

[4] Etacheri V, Di Valentin C, Schneider J, Bahnemann D and Pillai S C 2015 Visible-light activation of TiO_2 photocatalysts: Advances in theory and experiments *Journal of Photochemistry and Photobiology C: Photochemistry Reviews* **25** 1–29

[5] Bankovsky V, Bankovsky I, Bankovsky P, Isakova J, Djackova I, Sharipo A, Eskin J, Dišlers A,

Rozenstein R, Saricev V, Djacenko S, Makarenko V and Balodis U 2013 *Reverse–Spin® technology – innovative principle of microbial cultivation* (Riga, Latvia: Biosan)
[6] EUCAST 2019 *EUCAST reading guide for broth microdilution*
[7] EUCAST 2020 *Antimicrobial susceptibility testing EUCAST disk diffusion method version 8.0*
[8] Xie Y, He Y, Irwin P L, Jin T and Shi X 2011 Antibacterial activity and mechanism of action of zinc oxide nanoparticles against campylobacter jejuni *Appl. Environ. Microbiol.* **77** 2325
[9] Navale G R, Thripuranthaka M, Late D and Shinde S 2015 Antimicrobial activity of ZnO nanoparticles against pathogenic bacteria and fungi *JSM Nanotechnol Nanomed* **3** 1033
[10] Sharif S, Singh M, Kim S J and Schaefer J 2009 Staphylococcus aureus peptidoglycan tertiary structure from carbon-13 spin diffusion *J Am Chem Soc* **131** 7023–30
[11] Gumbart J C, Beeby M, Jensen G J and Roux B 2014 Escherichia coli peptidoglycan structure and mechanics as predicted by atomic-scale simulations *PLoS Comput Biol* **10** e1003475
[12] Hirota K, Sugimoto M, Kato M, Tsukagoshi K, Tanigawa T and Sugimoto H 2010 Preparation of zinc oxide ceramics with a sustainable antibacterial activity under dark conditions *Ceramics International* **36** 497–506
[13] Rago I, Chandraiahgari C R, Bracciale M P, Bellis G D, Zanni E, Guidi M C, Sali D, Broggi A, Palleschi C, Sarto M S and Uccelletti D 2014 Zinc oxide microrods and nanorods: different antibacterial activity and their mode of action against Gram-positive bacteria *RSC Adv.* **4** 56031–40

IOP Conf. Series: Materials Science and Engineering 1050 (2021) 012014 doi:10.1088/1757-899X/1050/1/012014

Anchor rods position assesment for connection of manipulating equipment to the transported structure

Š Vrtiel, M Naď and M Behúlová

Slovak University of Technology in Bratislava, Faculty of Materials Science and Technology in Trnava, Ulica Jána Bottu 2781/25, 917 24 Trnava, Slovak Republic

stefan.vrtiel@stuba.sk

Abstract. The manipulation and transport of baskets in which goods or components of various shapes are inserted is very often complicated due to manufacturing tolerances and deformations of the basket. Prediction of the connectivity of the basket and the transport device in the gripping points is very important from the point of view of safe transport of the basket. The methodology and procedure for determining the connectivity of the gripping elements of the transported device with the anchor holes of the basket columns is presented. In the analysis and evaluation of the connectivity of the basket with the transport device, the production tolerances of the basket and the transport device are taken into account and as well as the changes of important dimensions of the basket caused by different loading conditions.

1. Introduction

Manipulation and transport of goods or components with different shapes are very often performed by placing them in transport baskets. These baskets are usually connected to the manipulation device by various anchoring or gripping elements. It is evident that the manufacturing tolerances of the welded basket, as well as the deformation of the basket caused by the contents that are inserted into the basket, can significantly change the geometry of the basket. In addition, due to unexpected movement and the resulting change in the position of the basket contents, undesired deformation of the basket may occur. These geometrical changes in the basket shape, which would occur especially in places where the mutual anchoring of the basket to the manipulation device is expected, cause complications in their connection. As a result, it will not be possible to insert the gripping rods of the manipulation device into the anchor holes of the basket and the transport of the filled basket to the desired location will be impossible. This situation will caused absolutely unacceptable conditions in the case of transport of hazardous materials (for example, transport of radioactive waste - RAW), because the physical presence and activity leading to the correction of the condition that has occurred is absolutely excluded. In order to avoid these undesirable situations, detailed information on the manufacturing tolerances of the important basket dimensions and the gripping part of manipulation device have to be known. In addition, the some basket deformations are caused by predicted load cases, which are determined using stress-strain analyzes. These analyzes are performed on virtual computational models and resulting deformations have to be considered when the connectivity of the manipulation device to the basket is analyzed and evaluated. The procedure for analyzing and assessing the connectivity of the gripping parts of manipulation equipment to the transport basket for RAW materials is presented in this article.

2. Description of transport process

The manipulation device has four gripping rods (Figure 1) which are inserted into holes which have the shape of a "keyhole" and are located on the upper surface of the basket columns. After inserting the gripping rods into the holes of the "keyhole", these rods are rotated by 90 ° and a fixed connection of the basket with the manipulation device is achieved by means of transverse beams on the gripping rods [1].

Figure 1. Configuration of basket and manipulation device [1].

Based on the manufacturing tolerances of the basket and the gripping rods of the manipulation device and the deformations of the basket, a specific state can occur under certain conditions, which causes a mutual collision between the gripping rods and the "keyholes" on the basket columns. In this situation, it is not possible to insert and rotate the gripping rods and connect the basket to the manipulation device.

An important effect on the connectivity of the basket and the manipulation device is caused mainly by the deformation of the basket due to the action of loading conditions, which may occur due to improper or unstable arrangement of the contents inserted in the basket. The problem that can occur is caused, for example, by the so-called "domino effect" (Figure 2) and subsequent support of the fragments inserted in the basket on one of the columns of the basket. This causes a significant deformation of the basket and a change in the position of the "keyholes" and then the insertion of the bars of the manipulation device into these holes is not possible.

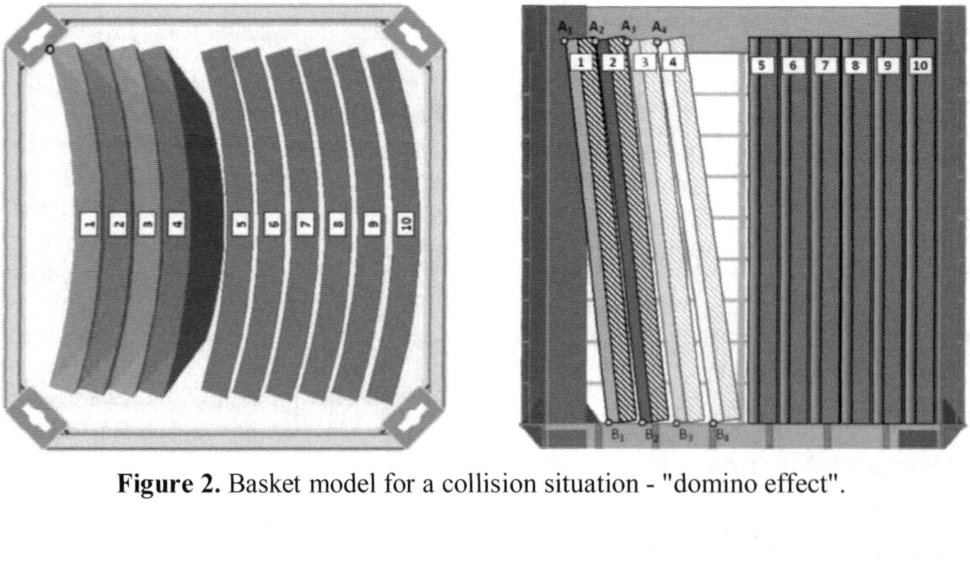

Figure 2. Basket model for a collision situation - "domino effect".

3. Analysis and assessment of connectivity of manipulation device and basket

The first fact that must be taken into account is the manufacturing tolerances of the basket, but also the manufacturing tolerances of the gripping rods of the handling equipment. From the actually made basket of the handling device, it is possible to determine by measurement the actual geometric configurations of the "keyholes" in the columns of the basket and the gripping rods. The coordinates of basket points A_b, B_b, C_b, D_b (Table 1) and coordinates of gripping rod points A_g, B_g, C_g, D_g (Table 2) are determined by measuring and they define the actual positions of the points.

The basket deformations (Figure 4a), which are determined by calculation using numerical simulations on a virtual basket model for the assumed load cases, change the positions of "keyholes" centers. The positions of the points A_b, B_b, C_b, D_b (Table 3) are changed to the position A_{bd}, B_{bd}, C_{bd}, D_{bd} (Table 4) in the coordinate system x_b and y_b (Figure 4b).

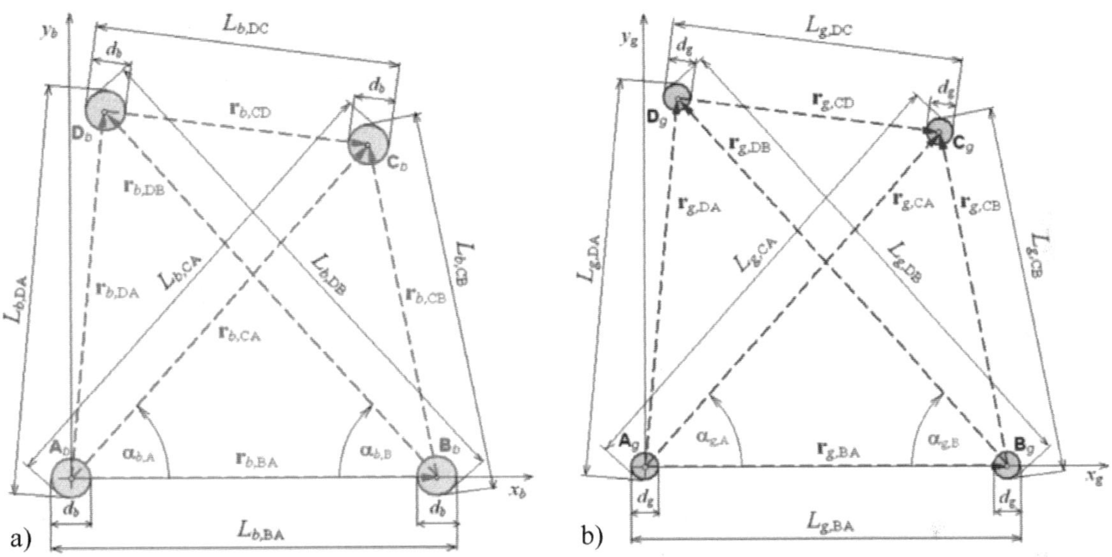

Figure 3. Model of "keyholes" of basket (a) and gripping rods (b) for determination of actual dimensions.

Table 1. Vector coordinates of basket points ("keyhole" center points - A_b, B_b, C_b, D_b).

$\mathbf{r}_{b,\mathrm{A}}$	$x_{b,\mathrm{A}} = 0$	$y_{b,\mathrm{A}} = 0$
$\mathbf{r}_{b,\mathrm{B}} = \mathbf{r}_{b,\mathrm{BA}}$	$x_{b,\mathrm{B}} = L_{b,BA} - d_b$	$y_{b\,k,\mathrm{B}} = 0$
$\mathbf{r}_{b,\mathrm{C}} = \mathbf{r}_{b,\mathrm{CA}}$	$x_{b,\mathrm{C}} = (L_{b,CA} - d_b)\cos\alpha_{b,A}$	$y_{b,\mathrm{C}} = (L_{b,CA} - d_b)\sin\alpha_{b,A}$
$\mathbf{r}_{b,\mathrm{D}} = \mathbf{r}_{b,\mathrm{DA}}$	$x_{b,\mathrm{D}} = L_{b,BA} - d_b - (L_{b,DB} - d_b)\cos\alpha_{b,B}$	$y_{b,\mathrm{D}} = (L_{b,DB} - d_b)\sin\alpha_{b,B}$
$\alpha_{b,A} = \arccos\left[\dfrac{(L_{b,CA} - d_b)^2 + (L_{b,BA} - d_b)^2 - (L_{b,CB} - d_b)^2}{2(L_{b,CA} - d_b)(L_{b,BA} - d_b)}\right]$		
$\alpha_{b,B} = \arccos\left[\dfrac{(L_{b,BA} - d_b)^2 + (L_{b,DB} - d_b)^2 - (L_{b,DA} - d_{bk})^2}{2(L_{b,BA} - d_b)(L_{b,DB} - d_b)}\right]$		

Table 2. Vector coordinates of gripping rods (center points of cross sections - A_g, B_g, C_g, D_g).

$\mathbf{r}_{g,A}$	$x_{g,A} = 0$	$y_{g,A} = 0$
$\mathbf{r}_{g,B} = \mathbf{r}_{g,BA}$	$x_{g,B} = L_{g,BA} - d_g$	$y_{g,B} = 0$
$\mathbf{r}_{g,C} = \mathbf{r}_{g,CA}$	$x_{g,C} = (L_{g,CA} - d_g)\cos\alpha_{g,A}$	$y_{g,C} = (L_{g,CA} - d_g)\sin\alpha g_{z,A}$
$\mathbf{r}_{g,D} = \mathbf{r}_{g,DA}$	$x_{g,D} = L_{g,BA} - d_g - (L_{g,DB} - d_g)\cos\alpha_{g,B}$	$y_{g,D} = (L_{g,DB} - d_g)\sin\alpha_{g,B}$
$\alpha_{g,A} = \arccos\left[\dfrac{(L_{g,CA}-d_g)^2 + (L_{g,BA}-d_g)^2 - (L_{g,CB}-d_g)^2}{2(L_{g,CA}-d_g)(L_{g,BA}-d_g)}\right]$		
$\alpha_{g,B} = \arccos\left[\dfrac{(L_{g,BA}-d_g)^2 + (L_{g,DB}-d_g)^2 - (L_{g,DA}-d_g)^2}{2(L_{g,BA}-d_g)(L_{g,DB}-d_g)} -\right]$		

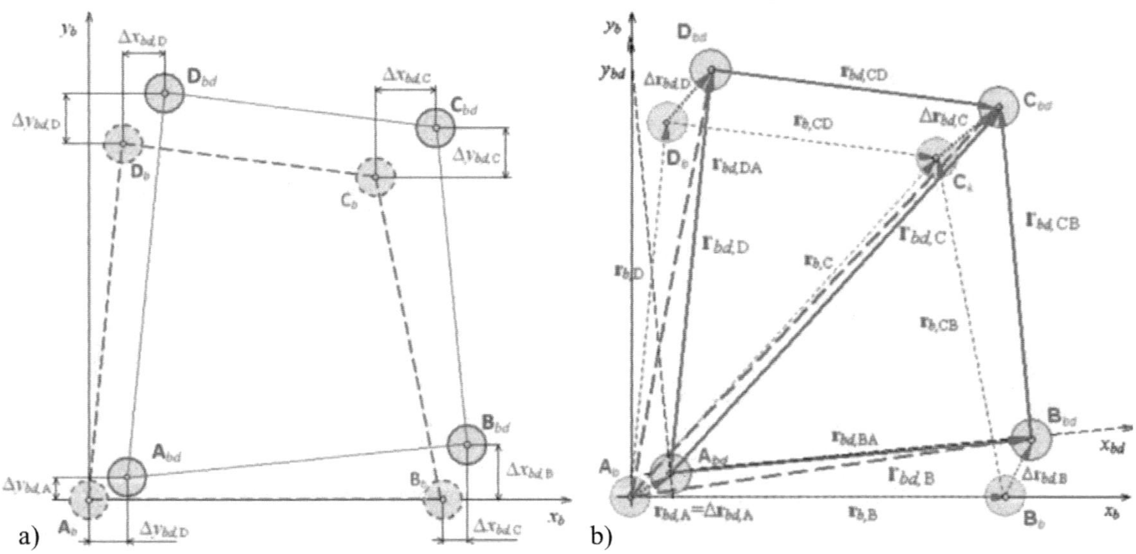

Figure 4. Change in the position of the basket points due to deformation.

Table 3. Position changes of points (A_b, B_b, C_b, D_b) caused by basket deformation (Figure 4a).

Displacement vectors	Point displacements (A_b, B_b, C_b, D_b)			
$\Delta\mathbf{r}_{bd,A} = \Delta x_{bd,A}\mathbf{i} + \Delta y_{bd,A}\mathbf{j}$	$\Delta x_{dk,A} =$	. . .	$\Delta y_{dk,A} =$	. . .
$\Delta\mathbf{r}_{bd,B} = \Delta x_{bd,B}\mathbf{i} + \Delta y_{bd,B}\mathbf{j}$	$\Delta x_{dk,B} =$	. . .	$\Delta y_{dk,B} =$	. . .
$\Delta\mathbf{r}_{bd,C} = \Delta x_{bd,C}\mathbf{i} + \Delta y_{bd,C}\mathbf{j}$	$\Delta x_{dk,C} =$	. . .	$\Delta y_{dk,C} =$	. . .
$\Delta\mathbf{r}_{bd,D} = \Delta x_{bd,D}\mathbf{i} + \Delta y_{bd,D}\mathbf{j}$	$\Delta x_{dk,D} =$	. . .	$\Delta y_{dk,D} =$	. . .

Table 4. Position vectors of points (A_{bd}, B_{bd}, C_{bd}, D_{bd}) after basket deformation.

Position vectors of points (A_{bd}, B_{bd}, C_{bd}, D_{bd}) to the coordinate system x_b, y_b (Figure 4b)		
$\mathbf{r}_{bd,A} = x_{bd,A}\mathbf{i} + y_{bd,A}\mathbf{j}$	$x_{bd,A} = \Delta x_{bd,A}$	$y_{bd,A} = \Delta y_{bd,A}$
$\mathbf{r}_{bd,B} = x_{bd,B}\mathbf{i} + y_{bd,B}\mathbf{j}$	$x_{bd,B} = x_{b,B} + \Delta x_{bd,B}$	$y_{bd,B} = \Delta y_{bd,B}$
$\mathbf{r}_{bd,C} = x_{bd,C}\mathbf{i} + y_{bd,C}\mathbf{j}$	$x_{bd,C} = x_{b,C} + \Delta x_{bd,C}$	$y_{bd,C} = y_{b,C} + \Delta y_{bd,C}$
$\mathbf{r}_{bd,D} = x_{bd,D}\mathbf{i} + y_{bd,D}\mathbf{j}$	$x_{bd,D} = x_{b,D} + \Delta x_{bd,D}$	$y_{bd,D} = y_{b,D} + \Delta y_{bd,D}$

To evaluate the mutual position of the corresponding points (A, B, C, D) of the basket and gripping rods (Table 5 and Table 6), it is necessary to express the position vectors for these points in the same coordinate system, i. e. to harmonize the coordinate system (x_{bd}, y_{bd}) (Figure 5a) with the coordinate system (x_g, y_g) (Figure 5b).

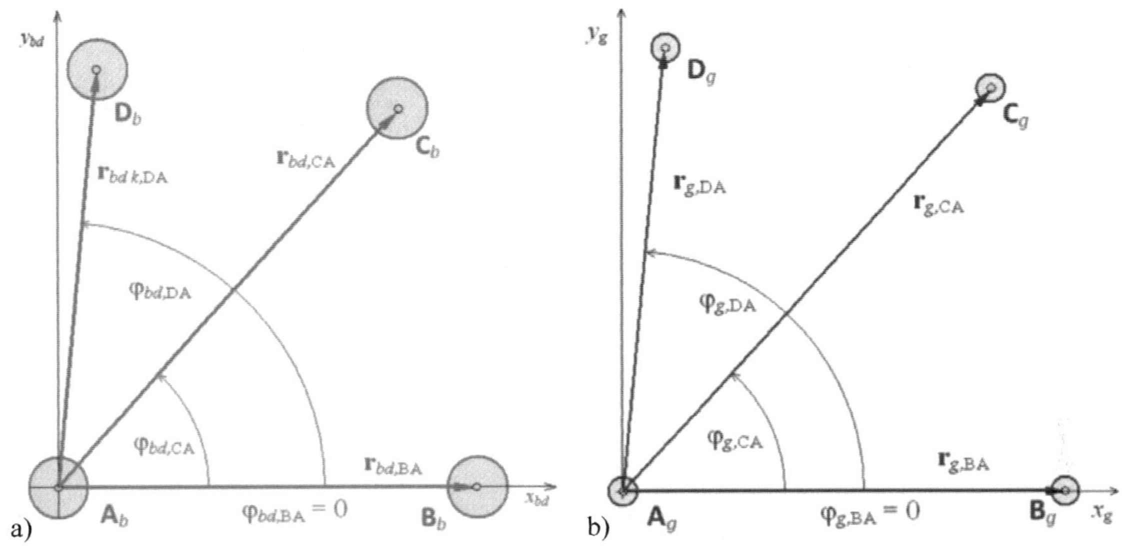

Figure 5. Models of coordinate systems (x_{bd}, y_{bd}) and (x_g, y_g) for their harmonization.

After harmonization of coordinate systems, it is possible to analyze the mutual position of gripping rods and "keyholes" in the basket columns (Figure 6). The mutual point positions of the gripping rods and the "keyholes" in the basket columns are defined by the coordinates (x_{gb}, y_{gb}) and by angles φ_{gb} (Table 7).

Table 5. Position of basket points (A_{bd}, B_{bd}, C_{bd}, D_{bd}) and angles to axis x_{bd} (Figure 5a).

Vectors defining the position of the basket points (x_{bd}, y_{bd})		
$\mathbf{r}_{bd,\mathrm{BA}} = x_{bd,\mathrm{BA}}\mathbf{i} + y_{bd,\mathrm{BA}}\mathbf{j}$	$x_{bd,\mathrm{BA}} = x_{bd,\mathrm{B}} - x_{bd,\mathrm{A}}$	$y_{bd,\mathrm{BA}} = y_{bd,\mathrm{B}} - y_{bd,\mathrm{A}}$
$\mathbf{r}_{bd,\mathrm{CA}} = x_{bd,\mathrm{CA}}\mathbf{i} + y_{bd,\mathrm{CA}}\mathbf{j}$	$x_{bd,\mathrm{CA}} = x_{bd,\mathrm{C}} - x_{bd,\mathrm{A}}$	$y_{bd,\mathrm{CA}} = y_{bd,\mathrm{C}} - y_{bd,\mathrm{A}}$
$\mathbf{r}_{bd,\mathrm{DA}} = x_{bd,\mathrm{DA}}\mathbf{i} + y_{bd,\mathrm{DA}}\mathbf{j}$	$x_{bd,\mathrm{DA}} = x_{bd,\mathrm{D}} - x_{bd,\mathrm{A}}$	$y_{bd,\mathrm{DA}} = y_{bd,\mathrm{D}} - y_{bd,\mathrm{A}}$
Angles of vectors $\mathbf{r}_{bd,\mathrm{BA}}$, $\mathbf{r}_{bd,\mathrm{CA}}$, $\mathbf{r}_{bd,\mathrm{DA}}$ to axis x_{bd}		
$\varphi_{bd,\mathrm{BA}}$	0	
$\varphi_{bd,\mathrm{CA}}$	$\varphi_{bd,\mathrm{CA}} = \arccos\left(\dfrac{\mathbf{r}_{bd,\mathrm{BA}} \cdot \mathbf{r}_{bd,\mathrm{CA}}}{\lvert \mathbf{r}_{bd,\mathrm{BA}} \rvert \cdot \lvert \mathbf{r}_{bd,\mathrm{CA}} \rvert}\right)$	
$\varphi_{bd,\mathrm{DA}}$	$\varphi_{bd,\mathrm{DA}} = \arccos\left(\dfrac{\mathbf{r}_{bd,\mathrm{BA}} \cdot \mathbf{r}_{bd,\mathrm{DA}}}{\lvert \mathbf{r}_{bd,\mathrm{BA}} \rvert \cdot \lvert \mathbf{r}_{bd,\mathrm{DA}} \rvert}\right)$	

Table 6. Position of gripping rods points (A_g, B_g, C_g, D_g) and angles to axis x_{bd} (Figure 5b).

Vectors defining the position of the gripping rods (x_g, y_g)		
$\mathbf{r}_{g,\text{BA}} = x_{g,\text{BA}}\mathbf{i} + y_{g,\text{BA}}\mathbf{j}$	$x_{g,\text{BA}} = x_{g,\text{B}} - x_{g,\text{A}}$	$y_{g,\text{BA}} = y_{g,\text{B}} - y_{g,\text{A}}$
$\mathbf{r}_{g,\text{CA}} = x_{g,\text{CA}}\mathbf{i} + y_{g,\text{CA}}\mathbf{j}$	$x_{g,\text{CA}} = x_{g,\text{C}} - x_{g,\text{A}}$	$y_{g,\text{CA}} = y_{g,\text{C}} - y_{g,\text{A}}$
$\mathbf{r}_{g,\text{DA}} = x_{g,\text{DA}}\mathbf{i} + y_{g,\text{DA}}\mathbf{j}$	$x_{g,\text{DA}} = x_{g,\text{D}} - x_{g,\text{A}}$	$y_{g,\text{DA}} = y_{g,\text{D}} - y_{g,\text{A}}$
Angles of vectors $\mathbf{r}_{g,\text{BA}}$, $\mathbf{r}_{g,\text{CA}}$, $\mathbf{r}_{g,\text{DA}}$ to axis x_g		
$\varphi_{g,\text{BA}}$	0	
$\varphi_{g,\text{CA}}$	$\varphi_{g,\text{CA}} = \arccos\left(\dfrac{\mathbf{r}_{g,\text{BA}} \cdot \mathbf{r}_{g,\text{CA}}}{\lvert \mathbf{r}_{g,\text{BA}} \rvert \cdot \lvert \mathbf{r}_{g,\text{CA}} \rvert}\right)$	
$\varphi_{g,\text{DA}}$	$\varphi_{g,\text{DA}} = \arccos\left(\dfrac{\mathbf{r}_{g,\text{BA}} \cdot \mathbf{r}_{g,\text{DA}}}{\lvert \mathbf{r}_{g,\text{BA}} \rvert \cdot \lvert \mathbf{r}_{g,\text{DA}} \rvert}\right)$	

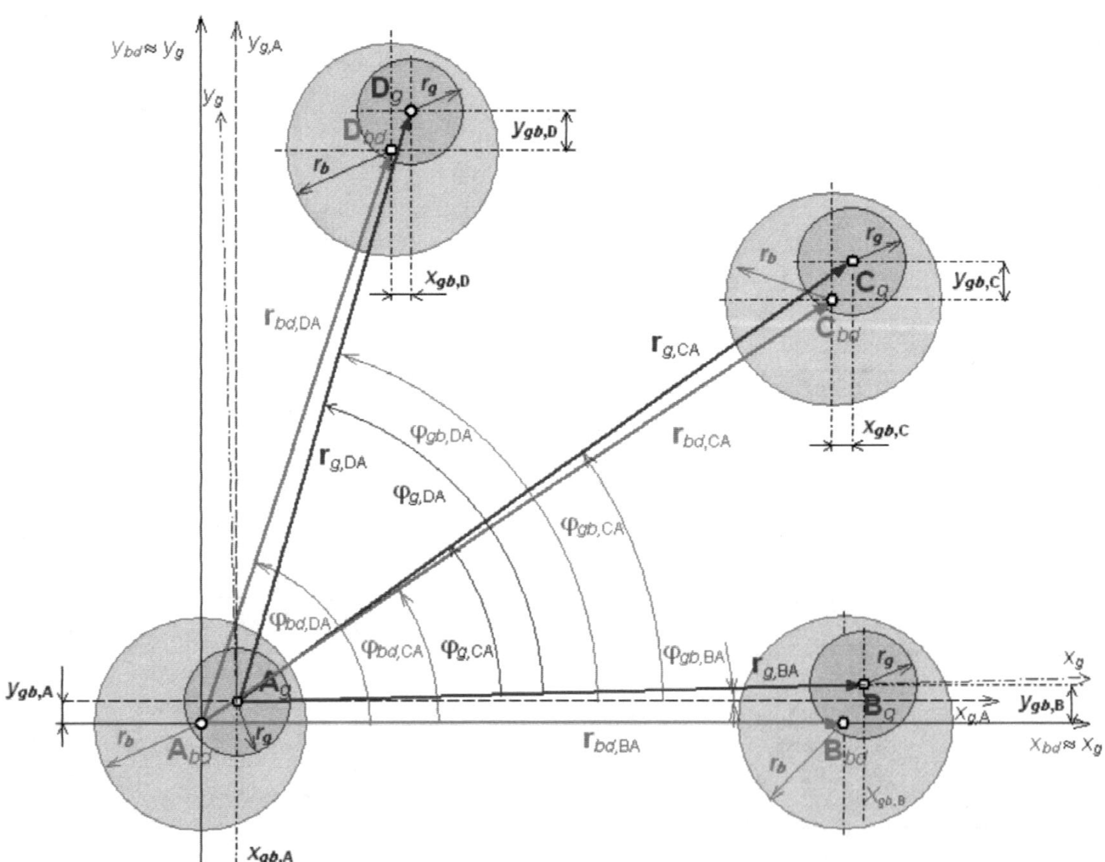

Figure 6. Harmonized coordinate systems (x_{bd}, y_{bd}) and (x_g, y_g).

Table 7. Position of gripping rods points A_g, B_g, C_g, D_g to basket points A_{bd}, B_{bd}, C_{bd}, D_{bd}.

Angles of vectors $\mathbf{r}_{g,BA}$, $\mathbf{r}_{g,CA}$, $\mathbf{r}_{g,DA}$ after displacements ($x_{gb,A}$, $y_{gb,A}$) of point A_g to point A_{bd} and after rotation of gripping rods against the basket by angle $\varphi_{gb,BA}$ to axis $x_{g,A}$	
$\varphi_{gb,BA}$ = "*set up*"	
$\varphi_{gb,CA} = \varphi_{g,CA} + \varphi_{gb,BA}$	
$\varphi_{gb,DA} = \varphi_{g,DA} + \varphi_{gb,BA}$	
Position of gripping rods points A_g, B_g, C_g, D_g to basket points A_{bd}, B_{bd}, C_{bd}, D_{bd}	
$x_{gb,A}$ = "*set up*"	$y_{gb,A}$ = "*set up*"
$x_{gb,B} = x_{gb,A} + \lvert \mathbf{r}_{g,BA} \rvert \cdot \cos(\varphi_{gb,BA}) - x_{bd,BA}$	$y_{gb,B} = y_{gb,A} + \lvert \mathbf{r}_{g,BA} \rvert \cdot \sin(\varphi_{gb,BA}) - y_{bd,BA}$
$x_{gb,C} = x_{gb,A} + \lvert \mathbf{r}_{g,CA} \rvert \cdot \cos(\varphi_{gb,CA}) - x_{bd,CA}$	$y_{gb,C} = y_{gb,A} + \lvert \mathbf{r}_{g,CA} \rvert \cdot \sin(\varphi_{gb,CA}) - y_{bd,CA}$
$x_{gb,D} = x_{gb,A} + \lvert \mathbf{r}_{g,DA} \rvert \cdot \cos(\varphi_{gb,DA}) - x_{bd,DA}$	$y_{gb,D} = y_{gb,A} + \lvert \mathbf{r}_{g,DA} \rvert \cdot \sin(\varphi_{gb,DA}) - y_{bd,DA}$

The collisions of the gripping rods and the "keyholes" of the basket columns (Figure 7) for the corresponding points (A, B, C, D) are determined and evaluated from the mutual relationship of the circles with the radii r_g and r_b.

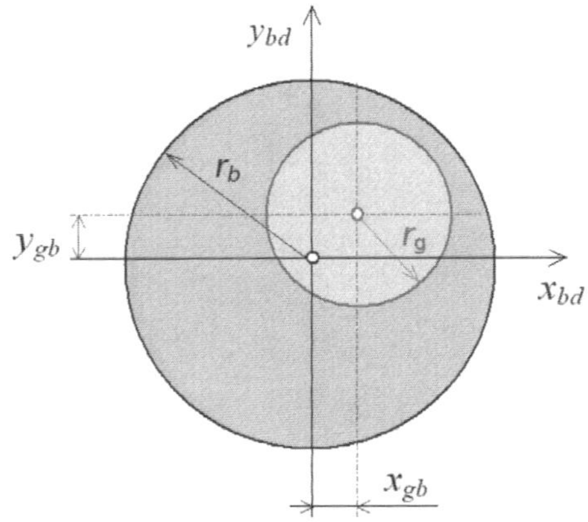

Figure 7. Model for assessment of collision of the gripping rod and the "keyhole".

The equations of the circles each point (A, B, C, D) with radii r_g and r_b to the local coordinate systems (x_{gb}, y_{gb}) are expressed by the following equations

$$x^2 + y^2 = r_b^2 \tag{1}$$

$$(x - x_{gb})^2 + (y - y_{gb})^2 = r_g^2 . \tag{2}$$

These circles can have three mutual positions:

1. circle with radius r_g is inside a circle with radius r_b - the circles have no common point,
2. circle with radius r_g is inside a circle with radius r_b and the circles touch at one point - the circles have one common point,
3. circle with radius r_g intersects a circle with radius r_b - the circles have two common points.

It is obvious that a collision state when gripping rods are inserted into the "keyholes" of column baskets does not occur if the positions of the circles at all points satisfy the first condition.

4. Conclusions

The methodology and procedure for the analysis and assessment of the connectivity of the manipulation device and the transport basket are presented. In assessing connectivity, the manufacturing tolerances of the gripping rods and the basket are taken into account, as well as the deformations of the basket caused by the predicted loading conditions. This procedure provides a suitable method for avoiding possible complications during the transport of hazardous materials such as RAW materials.

Acknowledgments

The research has been supported by the Scientific Grant Agency of the Slovak Republic within the Projects VEGA No. 1/0796/20 and KEGA No. 029STU-4/2018.

References

[1] Research study of the stress-strain state of the basket for the case of lateral loading of the basket column (Výskumná štúdia napäťovo-deformačného stavu koša pre prípad bočného zaťaženia stĺpika koša). Technical report - KAMM MTF STU 06/2020 (*In Slovak*).

AUTHOR INDEX